A Practical Guide to Quasi-elastic Neutron Scattering

A Practical Guide to Quasi-elastic Neutron Scattering

By

Mark T. F. Telling

Science and Technologies Facility Council, UK
Email: mark.telling@stfc.ac.uk

ROYAL SOCIETY
OF **CHEMISTRY**

Print ISBN: 978-1-78801-262-1
EPUB ISBN: 978-1-78801-926-2

A catalogue record for this book is available from the British Library

The Royal Society of Chemistry is a charity, registered in England and Wales, Number 207890, and a company incorporated in England by Royal Charter (Registered No. RC000524), registered office: Burlington House, Piccadilly, London W1J 0BA, UK, Telephone: +44 (0) 20 7437 8656.

Visit our website at www.rsc.org/books

Printed in the United Kingdom by CPI Group (UK) Ltd, Croydon, CR0 4YY, UK

For Dad

Preface

The technique of quasi-elastic neutron scattering (QENS) is a powerful experimental tool for extracting temporal and spatial information at the nanoscale from both soft and hard condensed matter systems. However, while seemingly simple, the method is beset with sensitivities that, if ill considered, can hinder data interpretation and possibly publication.

By highlighting key theoretical and data evaluation aspects of the technique, this specialised 'primer style' training resource encourages research success, and fruitful exploration, by guiding new researchers through a typical QENS experiment; from planning and sample preparation to data reduction and subsequent analysis.

While the book is designed to straddle existing resources and present a generic overview, there will be occasion where a specific neutron facility, research area or measurement method needs reference. The author also acknowledges that QENS measurements have been, are and will be possible on instruments and instrument classes not listed in this work. This in no way undervalues their contribution to the field.

Research examples are referenced throughout to illustrate the concepts addressed, with the book being written in such a way that it remains accessible to chemists, biologists, physicists and materials scientists. While not aimed specifically at undergraduates, the technique of neutron scattering is becoming embedded in core science,

A Practical Guide to Quasi-elastic Neutron Scattering
By Mark T. F. Telling
© Mark T. F. Telling 2020
Published by the Royal Society of Chemistry, www.rsc.org

as well as materials science, university syllabi. This text is therefore also viewed as a supplementary teaching aid when talking about the applicability of scattering methods in the chemistry, biology, materials science and physics classroom.

Mark T. F. Telling

Acknowledgements

This book could not have been completed without generous input and suggestion from the quasi-elastic neutron scattering community.

Sincere thanks to Valeria Arrighi, Ester Chiessi, Antonio Faraone, Sara Gabrielli, Barbara Gabrys, Victoria García Sakai, Olivia Kendall, Gaio Paradossi and Yosra Toumia for friendship and support during the writing of this book and for valued proof reading and/or feedback.

I would also like to thank, equally and in no particular order, the following for comment and input or for allowing me to reproduce their research in order to highlight the ideas outlined in this work:

Markus Appel, Marcella Cabrera Berg, Robert Bewley, Heloisa Bordallo, Irene Calvo Almazan, Suresh Mavila Chathoth, Robert Cywinski, Nicolas de Souza, Pascale Deen, Georg Ehlers, Hitoshi Endo, Fabrizia Foglia, Peter Fouquet, Bernhard Frick, Olaf Holderer, Mike Hore, Spencer Howells, Shinichi Itoh, Johanna Jochum, Fanni Juranyi, Toshi Kanaya, Yukinobu Kawakita, Mark Kibble, Oleg Kirichek, Wiebke Lohstroh, Eugene Mamontov, Michihiro Nagao, Kenji Nakajima, Francesca Natali, Kristine Niss, Colin Offer, Jacques Ollivier, Alexander O'Malley, Judith Peters, Shuo Qian, Arndt Remhof, Devinder Sivia, Andreas Stadler, Chris Stock, Xiao-Guang Sun, Jan Swenson, Madhusudan Tyagi, Andrew Wildes, Dehong Yu, Michaela Zamponi, Jean-Marc Zanotti, Dominik Zeller, Wei Zhou and Piotr Zolnierczuk.

Finally, to my family, wherever you may be, thank you and love always.

A Practical Guide to Quasi-elastic Neutron Scattering
By Mark T. F. Telling
© Mark T. F. Telling 2020
Published by the Royal Society of Chemistry, www.rsc.org

About The Author

Mark Telling is a senior research scientist who has over a decade's experience using, developing and advising on neutron methods, in particular the technique of quasi-elastic neutron scattering. His personal research interests focus on bio-materials. Mark is a visiting lecturer in the Department of Materials, University of Oxford and he has instructed students at workshops and research institutes across Europe, including the Universities of Rome (as Visiting Professor) and Copenhagen (at the Niels Bohr Institute). Mark has a comprehensive publication record as both first or co-author and he sits on several international research facility advisory panels. He is an active student mentor, a Science, Technology, Engineering and Mathematics (STEM) ambassador and a Trustee of the Institute of Physics.

A Practical Guide to Quasi-elastic Neutron Scattering
By Mark T. F. Telling
© Mark T. F. Telling 2020
Published by the Royal Society of Chemistry, www.rsc.org

Contents

A Practical Guide to Quasi-elastic Neutron Scattering
By Mark T. F. Telling
© Mark T. F. Telling 2020
Published by the Royal Society of Chemistry, www.rsc.org

6   Data Reduction   84

Part 3   Analysis   105

7   Elastic and Inelastic Fixed Window Scans   107

8 $S(Q,\omega)$ and $I(Q,t)$ 120

Part 1

Basics

1 If You Read Nothing Else...

In this chapter we will consider:

- Whether the QENS method is suitable for your research.
- Key parameters and preliminary experimental considerations necessary to plan a successful experiment.
- Research examples in the areas of chemistry, biology, physics and materials science.

1.1 "What can Quasi-elastic Neutron Scattering Do for Me?"

Since you have picked up this book it is probably safe to assume that you want to know, first and foremost, if the technique of quasi-elastic neutron scattering (QENS) can aid your research. To start, therefore, let's forget the where and how and simply ask:

"What can QENS do for me?"

Quasi-elastic neutron scattering (QENS) is a well-established experimental method for exploring low-energy collective (*i.e.* species move as an ensemble), or self (*i.e.* a specie moves alone), atomic fluctuations and/or molecular reorientation from both soft and hard condensed matter systems; dynamic phenomena that may be thought of as 'diffusive'. When we think of 'diffusive' here we might consider processes that are underpinned by atoms or molecules moving freely through a medium (*e.g.* the translational diffusion of molecules

A Practical Guide to Quasi-elastic Neutron Scattering
By Mark T. F. Telling
© Mark T. F. Telling 2020
Published by the Royal Society of Chemistry, www.rsc.org

in a bulk liquid) or alternatively along a geometrically constrained path (*e.g.* the restricted movement of side groups on a polymer backbone). For those interested in magnetism, the quasi-elastic scattering signal of interest might arise from magnetic fluctuations that are perhaps not harmonic and therefore do not occur at a well-defined frequency.

QENS experiments allow researchers to access motions that occur on picosecond (ps, 10^{-12} second) to nanosecond (ns, 10^{-9} second) timescales, and to explore said motions over length scales from 1 to ~500 Ångström (Å, where 1 Å = 1×10^{-10} m); length scales that cover both inter and intra molecular distances. Figure 1.1 illustrates the temporal and spatial ranges covered by the QENS technique relative to other experimental methods.

Numerous research fields have benefitted from a QENS investigation. A comprehensive overview is beyond the scope of this book. However, if we were to group themes under the broad headings of *Biology*, *Materials Science*, *Chemistry* and *Physics* then areas of study include:

Biology: protein structure–function-dynamics relationships;[1,2] phase behaviour of lipid bilayers and membrane shape fluctuations;[3,4] hydration shell[5] and cell crowding effects.[6]

Materials Science: hydrogen storage materials;[7] organic solar cell photo-voltaics;[8] built heritage;[9] nano-composites;[10] ion transport optimisation.[11]

Chemistry: ionic liquids;[12,13] surfactant mixtures;[14] micro emulsions;[15] physisorption;[16] polymer topology[17] and confinement effects;[18] proton conduction.[19]

Physics: spin fluctuations;[20] magnetic monopoles;[21] phonon lifetimes;[22] liquids under sheer and confinement;[23,24] theoretical predictions.[25]

In terms of the information extracted, and depending upon the nature of the investigation, QENS measurements allow researchers access to:

relaxation rates and associated distributions; bending moduli; immobile/mobile volume fractions; reaction kinetics; diffusion coefficients; transition temperatures; Debye–Waller factors; activation and binding energies; molecular rigidity and/or elastic compressibility; geometry and associated dimensionality of motion.

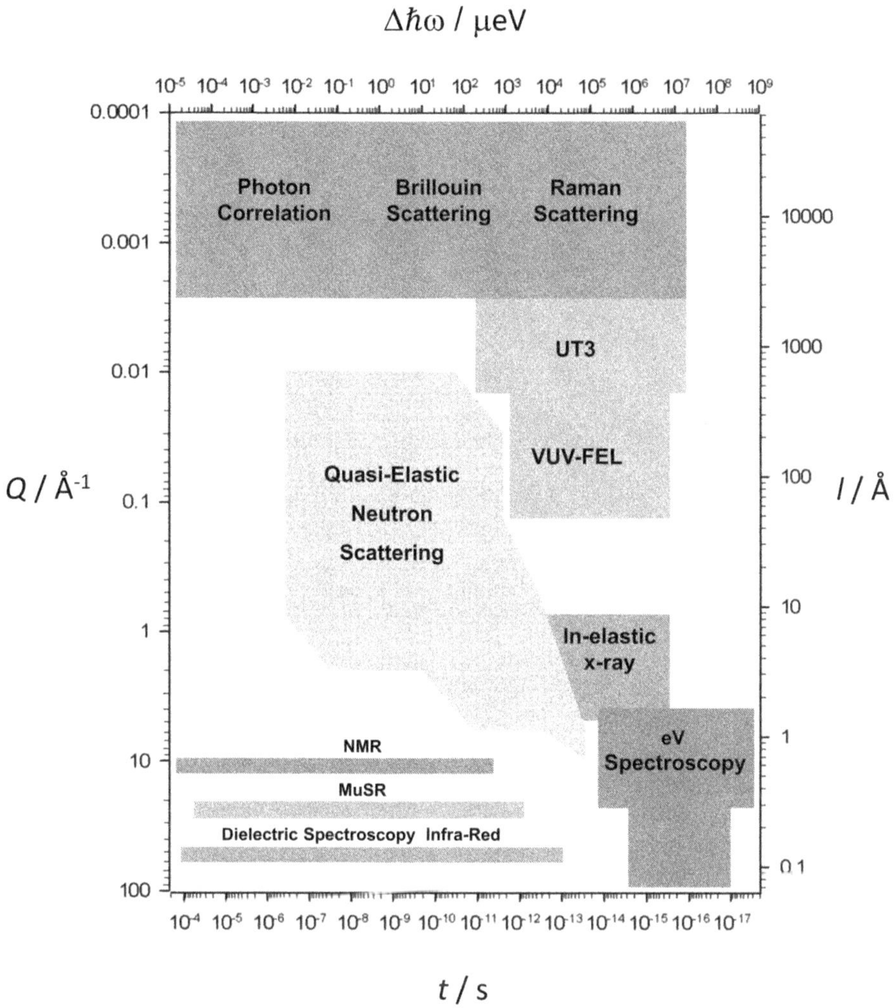

$\Delta\hbar\omega$ / µeV

Figure 1.1 Temporal and spatial ranges explored using quasi-elastic neutron scattering relative to other experimental methods. Figure
adapted from Telling.[26] The diagram correlates length (*l*) and
time (*t*). Those techniques that do not directly provide length
scale information are indicated as bars along the time axis.

To further accentuate research effort in these broad subject areas,
case studies from peer-reviewed works are summarized in Table 1.1.
These research examples expand upon the information that may be
obtained. Detailed information about each example can be found in
the accompanying reference.

Table 1.1   Research case studies.

Biology	*High hydrostatic pressure specifically affects molecular dynamics and shape of low-density lipoprotein particles.*[27]
Research aim	To study the effect of hydrostatic pressure (20–3000 bar) on the molecular dynamics, and shape, of low-density lipoprotein (LDL) particles; the physicochemical characteristics being relevant for proper functioning of lipid transport in blood circulation.
Why QENS?	To assign motions observed on different timescales to different dynamical populations, specific molecules or molecular groups within LDL and under pressure.
Additional characterisation/ supporting methods	• Small angle neutron scattering (SANS-II @ the Paul Scherrer Institut (PSI), Switzerland) • Sodium dodecyl sulfate polyacrylamide gel Electrophoresis (SDS-PAGE)
Main result reported	LDL copes well under high pressure conditions, although the lipid composition impacts the molecular dynamics and shape arrangement of the lipoprotein.
Instrument(s)/facility(ies) used and reported upper experimental observation time(s)	• IN5 (up to 100 ps) @ the Institut Laue–Langevin (ILL), France • IN6 (up to 15 ps) @ ILL, France • IN13 (up to 100 ps) @ ILL, France
Biology	*Photo-activation reduces side-chain dynamics of a light, oxygen and voltage (LOV) photoreceptor.*[28]
Research aim	To probe changes in conformational side-chain dynamics during photo-activation of flavin-binding light, oxygen and voltage (LOV) photoreceptors in both the dark and light states.
Why QENS?	To access the temporal range on which side chain dynamics are known to occur and to compare experimental data directly with molecular dynamics (MD) simulation.
Additional characterisation/ supporting methods	• Ultraviolet–visible spectroscopy (UV-Vis) • Molecular dynamics simulations (MD) • Dynamic light scattering (DLS)
Main result reported	Side-chain dynamics play a crucial role in photo-activation and signalling *via* modulation of conformational entropy.
Instrument(s)/facility(ies) used and reported upper experimental observation time(s)	• TOFTOF (up to 200 ps) operated by Technische Universität München (TUM) @ Heinz Maier-Leibnitz Zentrum (MLZ), Germany • SPHERES (up to 6 ns) operated by the Julich Centre for Neutron Science @ MLZ, Germany
Biology	*Neutron scattering studies of the interplay of amyloid β peptide and an anionic lipid 1,2-dimyristoyl-sn-glycero-3-phosphoglycerol.*[29]

Research aim	The interaction between lipid bilayers and amyloid β peptide (Aβ) plays a critical role in the proliferation of Alzheimer's disease. Deciphering the interplay of Aβ and lipid bilayers is of profound importance.
Why QENS?	To resolve *via* spectral analysis two different kinds of motion: the lateral and internal motions of the lipid molecules within the monolayer and the internal motion of the lipid molecules themselves.
Additional characterisation/ supporting methods	• SANS (Bio-SANS @ the High-Flux Isotope Reactor (HFIR), Oak Ridge National Laboratory (ORNL), USA and EQ-SANS @ the Spallation Neutron Source (SNS), ORNL, USA) • Circular dichroism (CD) • Fourier-transform infrared spectroscopy (FTIR) • Neutron membrane diffraction (NMD)
Main result reported	The addition of Aβ mainly affects the slower lateral motion of lipid molecules, but not the faster internal motion. The results reveal that Aβ associates with the highly charged membrane surface with limited impact on the structure. The altered membrane dynamics, however, could have influence on other membrane processes.
Instrument(s)/facility(ies) used and reported upper experimental observation time(s)	• BASIS (up to 400 ps) @ SNS, USA

Materials Science	*Nanoscale mobility of aqueous poly-acrylic acid in dental restorative cements.*[30]
Research aim	To improve the strength and longevity of glass ionomer cements used in dental care.
Why QENS?	To extract information about solvent mobility and hydrogen bonding kinetics as a function of solvent type and cement porosity.
Additional characterisation/ supporting methods	• Thermogravimetric analysis (TGA) • Fourier transform infrared spectroscopy (FTIR) • Inelastic neutron spectroscopy (VISION @ SNS, ORNL, USA)
Main result reported	The evolution of solvent mobility, water-polymer binding, polymer cross-linking and material density changes are all key considerations when developing improved dental restorative materials.
Instrument(s)/facility(ies) used and reported upper experimental observation time(s)	• PELICAN (up to 60 ps) @ Australian Nuclear Science and Technology Organisation (ANSTO), Australia • IN16B (up to 5 ns) @ ILL, France • IRIS (up to 200 ps) @ the ISIS Pulsed Neutron and Muon Source (ISIS), UK

Materials Science	*Quasi-elastic neutron scattering study of microscopic dynamics in polybutadiene reinforced with an unsaturated carboxylate.*[31,32]

(*continued*)

Table 1.1 (continued)

Research aim	To understand the superior mechanical properties of polybutadiene rubber (BR) reinforced with zinc di-acrylate (ZDA), compared to BR reinforced with carbon black or silica, *via* consideration of the effect of ZDA concentration on polymer dynamics.
Why QENS?	To observe material 'stiffness' *via* the mean squared displacement parameter and extract mobile volume fractions on the timescale of interest.
Additional characterisation/ supporting methods	• Differential scanning calorimetry (DSC) • Small-angle X-ray scattering (SAXS) • Small-angle neutron scattering (SANS)
Main result reported	The network-like structure of the BR, having a high crosslinking density around ZDA aggregates, is principally responsible for the high elastic modulus of ZDA/BR.
Instrument(s)/facility(ies) used and reported upper experimental observation time(s)	• DNA (up to 1 ns) @ the Materials and Life Science Experimental Facility (MLF), J-PARC, Japan
Materials Science	*Ion dynamics in ionic-liquid-based Li-ion electrolytes investigated by neutron scattering and dielectric spectroscopy.*[33]
Research aim	To understand the diffusion mechanisms of ions in pure and doped ionic liquids for the design of new ionic-liquid based energy storage media. In particular, the relationship between viscosity, ionic conductivity, diffusion coefficients and re-orientational dynamics in imidazolium-based ionic material.
Why QENS?	To detect directly ion mobility and relate it to ion transport mechanisms *via* analysis and separation of coherent (SNS-NSE) and incoherent scattering (BASIS).
Additional characterisation/ supporting methods	• Differential scanning calorimetry (DSC) • Dielectric spectroscopy (DS)
Main result reported	Two diffusion mechanisms are observed (translational and localised). Ion transport, along with ion aggregates, play a significant role in the ion diffusion process.
Instrument(s)/facility(ies) used and reported upper experimental observation time(s)	• SNS-NSE (up to 30 ns) @ SNS, USA • BASIS (up to 4 ns) @ SNS, USA
Chemistry	*Ammonia mobility in chabazite: insight into the diffusion component of the NH3-SCR process.*[34]
Research aim	To investigate the effect of counter ions on selective catalytic reduction.

Why QENS?	To extract diffusion information about ammonia across a commercial copper-chabazite zeolite catalyst and compare the results directly with molecular dynamics simulations.
Additional characterisation/ supporting methods	• Molecular dynamics simulations (MD)
Main result reported	The presence of counter ions has little impact on the diffusion of ammonia. Molecular dynamics simulations show that strong coordination of NH_3 with Cu^{2+} counter ions in the centre of the chabazite cage shield other molecules from interaction with the ion, allowing diffusion through the windows of the chabazite structure to continue freely.
Instrument(s)/facility(ies) used and reported upper experimental observation time(s)	• OSIRIS (up to 150 ps) @ ISIS, UK
Chemistry	*Benzene diffusion on graphite described by a rough hard disc model.*[25]
Research aim	To test existing models of friction and diffusion at the nanoscale using benzene molecules adsorbed on a graphite surface.
Why QENS?	Allows the nanometre motions of the benzene molecules to be followed from the picosecond to the nanosecond. In addition, MD simulations allow both the coherent and incoherent dynamic structure factors, $S_c(\boldsymbol{Q},\omega)$ and $S_i(\boldsymbol{Q},\omega)$, to be calculated in order to separate out, and understand, contributions from molecule–molecule and molecule–substrate interactions.
Additional characterisation/ supporting methods	• BET (Brunauer–Emmett–Teller) adsorption isotherms • Molecular dynamics simulation (MD)
Main result reported	The strong coverage dependence of the benzene diffusion demonstrates that diffusivity at the nanometre is driven by molecule–molecule interactions.
Instrument(s)/facility(ies) used and reported upper experimental observation time(s)	• IN11C (up to 4 ns) @ ILL, France • OSIRIS (up to 100 ps) @ ISIS, UK • IN6 (up to 50 ps) @ ILL, France
Chemistry	*Local structure and relaxation dynamics in the brush of polymer grafted silica nanoparticles.*[35]
Research aim	To investigate local structure and relaxation dynamics in the brush region of polymer-grafted silica nanoparticles.
Why QENS?	To obtain a direct measurement of both local structure and relaxation dynamics of concentrated, and semi-dilute, polymer brush regions.

(continued)

Table 1.1 (continued)

Additional characterisation/ supporting methods	• Activated CPDB *via* ^{1}H-NMR • Ultraviolet–visible spectroscopy (UV-Vis) • Size exclusion chromatography (SEC-MALS) • Small-angle neutron scattering (SANS)
Main result reported	Near the nanoparticle surface, a higher polymer chain concentration leads to stretched chains and slower dynamics compared to the response observed away from the surface.
Instrument(s)/facility(ies) used and reported upper experimental observation time(s)	• CHRNS – NSE (up to 40 ns) @ the National Institute of Standards and Technology (NIST) Centre for Neutron Research, USA

Physics	*Generalised spin-glass relaxation.*[20]
Research aim	To investigate the origins and implications of non-exponential relaxation in spin glasses, test theoretical models and better understand dynamical correlations associated with the onset of a glassy state.
Why QENS?	To follow non-exponential relaxation processes over several tens of nanoseconds, and extended length scales, in a single measurement.
Additional characterisation/ supporting methods	• X-ray powder diffraction (XRD) • AC susceptibility (χ_{ac})
Main result reported	Spin relaxation close to the glass temperature of CuMn and AuFe spin glasses is shown to follow a generalised exponential function that explicitly introduces hierarchically constrained dynamics and macroscopic interactions.
Instrument(s)/facility(ies) used and reported upper experimental observation time(s)	• IN11 (up to 40 ns) @ ILL, France

Physics	*Microscopic origin of the logarithmic relaxation in molecular glass-forming liquids.*[36]
Research aim	Logarithmic relaxation is a unique relaxation process exhibited by a few molecular liquids and biomolecules. However, the microscopic origin of logarithmic relaxation is unclear.
Why QENS?	The timescale of QENS is appropriate to probe fast β-relaxation processes. The neutron method also affords the possibility of using deuterated solvents to isolate specific motion(s).
Additional characterisation/ supporting methods	• Dynamic light scattering (DLS) and depolarised DLS

Main result reported	The results indicate that the intermolecular potential does not play a role in determining the logarithmic relaxation process. Instead, the coupling of localised and translational relaxation processes is a candidate for the origin of the logarithmic relaxation process exhibited by the molecular liquids.
Instrument(s)/facility(ies) used and reported upper experimental observation time(s)	• MARS (up to 150 ps) @ The Swiss Spallation Neutron Source (SINQ), PSI, Switzerland • PELICAN (up to 60 ps) @ ANSTO, Australia

Physics	*Interplay between static and dynamic polar correlations in relaxor Pb(Mg$_{1/3}$Nb$_{2/3}$)O$_3$.*[37]
Research aim	To establish whether the spatially localised regions of polar order in highly piezoelectric Pb(Mg$_{1/3}$Nb$_{2/3}$)O$_3$ (PMN) are static or dynamic. If the latter, to probe the timescale of the associated fluctuations.
Why QENS?	Since the fluctuations are not harmonic, and therefore do not occur at a finite frequency, it is prudent to measure quasi-elastic excitations around the elastic line. Triple-axis, backscattering and spin echo measurements are needed to explore a wide dynamic range.
Additional characterisation/ supporting methods	• Extensive elastic measurements, using time-of-flight and also diffraction, to determine where in momentum space the scattering cross section is maximum • Measurement of acoustic phonons to isolate the low energy dynamic response
Main result reported	That the spatially localised regions of polar order are dynamic.
Instrument(s)/facility(ies) used and reported upper experimental observation time(s)	• IN10 (up to 1 ns) @ ILL, France • IN11 (up to 40 ns) @ ILL, France

To explore whether the QENS method might match your particular research need in terms of parameter space, Table 1.2 outlines the experimental extremes associated with the technique, as well as key pre-experiment considerations necessary to plan a successful neutron scattering study. Conversion factors that will help you align the reported units with those extracted using different experimental methods are given in Appendix 1. Of course, the limits quoted are operating bounds. No single neutron instrument, or set of instruments

Table 1.2 Key parameters, operating limits and pre-experiment considerations necessary to plan a successful QENS study.

	Minimum	Maximum
Temporal range	~0.001 nanosecond (ns)	~1000 nanoseconds (ns)
Spatial range	~1 Ångström (Å)	~500 Ångström (Å)
Sample mass	~100 milligrams (mg)	~Grams (g)
	Ideally enough material to fully cover a surface area of a few cm² to match the incident neutron beam profile and thus optimise use of available neutron flux.	See Minimum. In practice, the upper limit is usually governed by neutron transmission through the sample; transmission being set at 85–90% to 'avoid' multiple scattering correction (see Part 2).
Temperature range	~50 milliKelvin (mK)- Using dilution refrigeration	~2273 Kelvin (K) Using furnace apparatus
Pressure range	0 kilobar (kbar)	• High pressure gas cells: up to 9 kbar (0.9 GPa) • Clamped cells: up to 20 kbar (2 GPa) • Paris Edinburgh systems: up to 100 kbar (10 GPa) • On specialized instruments up to 30 GPa
External magnetic field	0 Tesla (T)	• Superconducting magnets up to 15 T • Pulsed magnetic fields up to 40 T
Key experimental considerations	Ensure that the relaxation rate to be probed aligns with the experimental observation time afforded by QENS instrumentation, and that said motion occurs on an appropriate length scale. QENS studies investigate either collective (*i.e.* how atoms or molecules move as an ensemble), or self (*i.e.* how atoms or molecules move alone), motion(s). The former requires a coherent scattering response, the latter an incoherent scattering response (see Chapter 2). Guided by the motion type you wish to investigate, ensure the elemental composition of your sample exhibits a corresponding, and sizeable, associated scattering contribution. Neutrons can also be absorbed (rather than scattered) by some elements *e.g.* cadmium, gadolinium, lithium. Ensure that your sample does not have a high neutron absorbing composition. Sample containers are mostly constructed from aluminum. Check that the sample is non-corrosive. Other container materials, or anodization, can be used but may require bespoke machining and/or induce a potentially problematic background response.	
Possible ancillary options	While not routine, some neutron instruments offer additional *in situ* measurement methods *e.g.* Raman, calorimetry, humidity.	

at a neutron science institute (aka neutron facility), can access these ranges in their entirety. Nonetheless, Table 1.2 should provide suitable guidance when considering whether or not the parameter space covered by the QENS method spans that of your particular scientific problem.

1.2 The Importance of Complementary Information

From the case studies presented it is clear that QENS experiments are rarely performed alone. Instead they are underpinned by detailed pre-experiment sample characterisation, purity checks and supported with pre- or post-experiment auxiliary information. The former is particularly relevant if one wants to ensure that the material under investigation is phase pure, suitable for the scattering of neutrons and that the 'diffusive' component(s) of interest migrates on a time and length scale compatible with the quasi-elastic method. The main pre- and post-characterisation methods referred to in the above case studies are listed in Appendix 2 alongside brief descriptions of the supporting information they provide. At the very least, before a QENS study is attempted, it is sensible to understand your sample in terms of:

- Structure:
  - *e.g.* does the sample undergo a structural transition? If so, at what temperature/pressure/magnetic field *etc.?*
- Physical appearance, malleability, reactivity and specific handling issues:
  - Such information will guide, for example, the choice of sample container and the possible need for specific laboratory apparatus.
- Thermodynamic properties:
  - *e.g.* does the material show bulk phase change behaviour? If so, under what conditions?
- Complexity:
  - *e.g.* is the material single or multi-component? If so, what are the relative phase fractions? Interpreting quasi-elastic scattering from complex materials may prove challenging.
- Ease of selective labelling (aka deuteration).
  - *i.e.* can the ^1H atoms be replaced with deuterium in order to 'isolate' specific molecular species (see Chapter 2)? If so, can the deuteration procedure be completed on a reasonable timescale, at a reasonable cost and produce the minimum sample mass required for a quasi-elastic neutron scattering study?

Hopefully, an informed decision can now be made regarding the application of the quasi-elastic neutron scattering method to your particular research interest. If a QENS study seems prudent then let's proceed and consider the basics of this experimental technique.

References

1. L. Hong, N. Jain, X. Cheng, A. Bernal, M. Tyagi and J. C. Smith, *Sci. Adv.*, 2016, **2**, e1600886.
2. M. Grimaldo, F. Roosen-Runge, F. Zhang, F. Schreiber and T. Seydel, *Q. Rev. Biophys.*, 2019, **52**, e7.
3. S. Gupta, J. U. De Mel and G. J. Schneider, *Curr. Opin. Colloid Interface Sci.*, 2019, **42**, 121–136.
4. C. Monzel and K. Sengupta, *J. Phys. D: Appl. Phys.*, 2016, **49**, 243002.
5. A. Frölich, F. Gabel, M. Jasnin, U. Lehnert, D. Oesterhelt, A. M. Stadler, M. Tehei, M. Weik, K. Wood and G. Zaccai, *Faraday Discuss.*, 2009, **141**, 117–130.
6. S. Longeville and L.-R. Stingaciu, *Sci. Rep.*, 2017, **7**, 10448.
7. E. Mamontov, A. I. Kolesnikov, S. Sampath and J. L. Yarger, *Sci. Rep.*, 2017, **7**, 16244.
8. T. Etampawala, D. Ratnaweera, B. Morgan, S. Diallo, E. Mamontov and M. Dadmun, *Polymer*, 2015, **61**, 155–162.
9. J. Jacobsen, M. S. Rodrigues, M. T. F. Telling, A. L. Beraldo, S. F. Santos, L. P. Aldridge and H. N. Bordallo, *Sci. Rep.*, 2013, **3**, 2667.
10. D. Bhowmik, J. A. Pomposo, F. Juranyi, V. García-Sakai, M. Zamponi, Y. Su, A. Arbe and J. Colmenero, *Macromolecules*, 2014, **47**, 304–315.
11. X. C. Chen, R. L. Sacci, N. C. Osti, M. Tyagi, Y. Wang, M. J. Palmer and N. J. Dudney, *Mol. Syst. Des. Eng.*, 2019, **4**, 379–385.
12. S. Liu, C. Liedel, N. V. Tarakina, N. C. Osti and P. Akcora, *Nanoscale*, 2019, **11**, 19832.
13. T. Burankova, J. F. Mora Cardozo, D. Rauber, A. Wildes and J. P. Embs, *Sci. Rep.*, 2018, **8**, 16400.
14. I. Hoffmann, M. Simon, B. Farago, R. Schweins, P. Falus, O. Holderer and M. Gradzielski, *J. Chem. Phys.*, 2016, **145**, 124901.
15. V. K. Sharma, D. G. Hayes, S. Gupta, V. S. Urban, H. M. O'Neill, S. V. Pingali, M. Ohl and E. Mamontov, *J. Phys. Chem. C*, 2019, **123**, 11197–11206.
16. O. Czakkel, B. Nagy, G. Dobos, P. Fouquet, E. Bahn and K. László, *Int. J. Hydrogen Energy*, 2019, **44**, 18169–18178.
17. V. Arrighi, S. Gagliardi, F. Ganazzoli, J. S. Higgins, G. Raffaini, J. Tanchawanich, J. Taylor and M. T. F. Telling, *Macromolecules*, 2018, **51**, 7209–7223.
18. D. Richter and M. Kruteva, *Soft Matter*, 2019, **15**, 7316–7349.
19. A. Braun and Q. Chen, *Nat. Commun.*, 2017, **8**, 15830.
20. R. M. Pickup, R. Cywinski, C. Pappas, B. Farago and P. Fouquet, *Phys. Rev. Lett.*, 2009, **102**, 097202.
21. S. Gao, O. Zaharko, V. Tsurkan, L. Prodan, E. Riordan, J. Lago, B. Fåk, A. R. Wildes, M. M. Koza, C. Ritter, P. Fouquet, L. Keller, E. Canévet, M. Medarde, J. Blomgren, C. Johansson, S. R. Giblin, S. Vrtnik, J. Luzar, A. Loidl, C. Rüegg and T. Fennell, *Phys. Rev. Lett.*, 2018, **120**, 137201.
22. P.-F. Lory, S. Pailhès, V. M. Giordano, H. Euchner, H. D. Nguyen, R. Ramlau, H. Borrmann, M. Schmidt, M. Baitinger, M. Ikeda, P. Tomeš, M. Mihalkovič, C. Allio,

M. R. Johnson, H. Schober, Y. Sidis, F. Bourdarot, L. P. Regnault, J. Ollivier, S. Paschen, Y. Grin and M. de Boissieu, *Nat. Commun.*, 2017, **8**, 491.

23. C. Bousige, S. Rols, J. Ollivier, H. Schober, P. Fouquet, G. G. Simeoni, V. Agafonov, V. Davydov, Y. Niimi, K. Suenaga, H. Kataura and P. Launois, *Phys. Rev. B*, 2013, **87**, 195438.

24. G. Briganti, G. Rogati, A. Parmentier, M. Maccarini and F. De Luca, *Sci. Rep.*, 2017, **7**, 45021.

25. I. Calvo-Almazan, E. Bahn, M. M. Koza, M. Zbiri, M. Maccarini, M. T. F. Telling, S. Miret-Artes and P. Fouquet, *Carbon*, 2014, **79**, 183–191.

26. M. T. F. Telling, in *Dynamics of Biological Macromolecules by Neutron Scattering*, ed. S. Magazù and F. Migliardo, Bentham Science Publishers, 2011, p. 4.

27. M. Golub, B. Lehofer, N. Martinez, J. Ollivier, J. Kohlbrecher, R. Prassl and J. Peters, *Sci. Rep.*, 2017, **7**, 46034.

28. A. M. Stadler, E. Knieps-Grunhagen, M. Bocola, W. Lohstroh, M. Zamponi and U. Krauss, *Biophys. J.*, 2016, **110**, 1064–1074.

29. D. K. Rai, V. K. Sharma, D. Anunciado, H. O'Neill, E. Mamontov, V. Urban, W. T. Heller and S. Qian, *Sci. Rep.*, 2016, **6**, 30983.

30. M. C. Berg, A. R. Benetti, M. T. F. Telling, T. Seydel, D. H. Yu, L. L. Daemen and H. N. Bordallo, *ACS Appl. Mater. Interfaces*, 2018, **10**, 9904–9915.

31. R. Mashita, R. Inoue, T. Tominaga, K. Shibata, H. Kishimoto and T. Kanaya, *Soft Matter*, 2017, **13**, 7862–7869.

32. R. Mashita, H. Kishimoto, R. Inoue, A. Koda, R. Kadono and T. Kanaya, *Polymer*, 2016, **105**, 510–515.

33. C. J. Jafta, C. Bridges, L. Haupt, C. Do, P. Sippel, M. J. Cochran, S. Krohns, M. Ohl, A. Loidl, E. Mamontov, P. Lunkenheimer, S. Dai and X. G. Sun, *Chemsuschem*, 2018, **11**, 3512–3523.

34. A. J. O'Malley, I. Hitchcock, M. Sarwar, I. P. Silverwood, S. Hindocha, C. R. A. Catlow, A. P. E. York and P. J. Collier, *Phys. Chem. Chem. Phys.*, 2016, **18**, 17159–17168.

35. Y. Wei, Y. F. Xu, A. Faraone and M. J. A. Hore, *ACS Macro Lett.*, 2018, **7**, 699–704.

36. C. J. Chen, R. P. Krishnan, K. K. Wong, D. H. Yu, F. Juranyi and S. M. Chathoth, *Phys. Rev. B*, 2018, **98**, 094203.

37. C. Stock, L. Van Eijck, P. Fouquet, M. Maccarini, P. M. Gehring, Guangyong Xu, H. Luo, X. Zhao, J.-F. Li and D. Viehland, *Phys. Rev. B*, 2010, **81**, 144127.

2 What Is QENS?

In this chapter we will consider:

- Key neutron scattering concepts.
- The benefits and limitations of using the neutron as a probe of matter.
- The roles that coherent and incoherent neutron scattering play in the identification of collective and self-motions.

2.1 Scattering Concepts

Experimentally, one can think of a QENS data set as either a series of (i) peaks of detected neutron intensity, I, and width Γ (usually quoted in terms of its full width at half maximum (FWHM) and defined in units of energy) or (ii) decay curves described in units of time (Figure 2.1). Peak intensity, width, or relaxation rate can exhibit marked angular variation, *i.e.* from one neutron detector to the next, and it is by modelling this variation that information about the frequency, and possibly geometry, of the process under investigation is extracted. As we will see in Chapter 3, the type of response recorded depends upon the type of neutron instrument used but, of course, both behaviours are related; a response in time being the Fourier transform of a spectrum measured in energy.

However, before we start interpreting our quasi-elastic scattering data sets, we first need to consider the concepts that underpin the neutron method, and the themes that recur frequently in the scientific

A Practical Guide to Quasi-elastic Neutron Scattering
By Mark T. F. Telling
© Mark T. F. Telling 2020
Published by the Royal Society of Chemistry, www.rsc.org

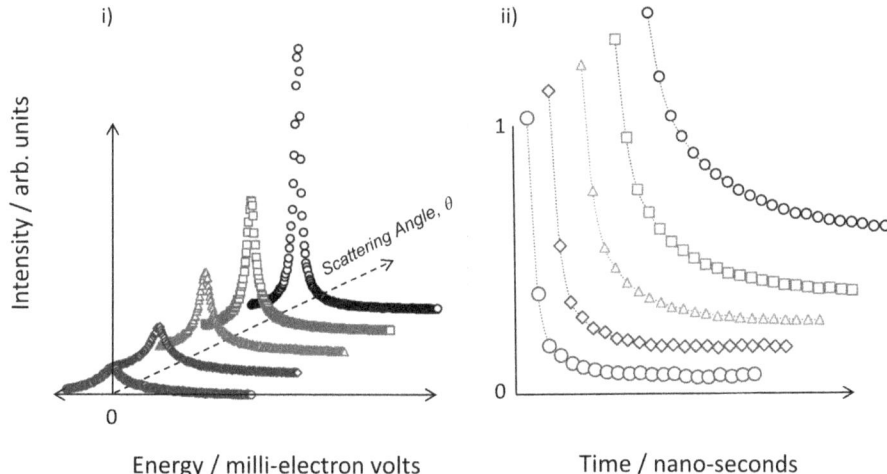

Figure 2.1 Illustrative QENS spectra as recorded in different neutron detectors and as a function of decreasing scattering angle. (i) Response in the energy domain. (ii) Response in the time domain. Conversion from (i) to (ii) can be achieved *via* a Fourier transform. The data shown in (i) is the so-called dynamic incoherent structure factor, *i.e.* $S_i(\mathbf{Q},\omega)$, and (ii) the associated normalised self-intermediate scattering function, *i.e.* $I_s(\mathbf{Q},t)$, from a linear polymer at 298 K.[1]

literature on neutron science. We will therefore use this chapter, and a predominantly semi-classical approach, to build a phenomenological understanding of:

- the scattering cross section
- coherent and incoherent scattering
- momentum transfer
- the double differential scattering cross section
- energy transfer (elastic and in-elastic scattering)
- selective deuteration.

But first, let's ask ourselves …

2.2 Why Use Neutrons at All?

As with any scattering method, the wavelength or energy of the incident radiation needs to be comparable to the length scale and/or excitation frequency probed. Light and X-rays are photonic in nature and demonstrate wave-like properties. The wavelength of light is of the order of a few thousand Ångström (λ_{light} ~4000–7000 Å) with $\lambda_{\text{X-ray}}$ ~0.01 to 100 Å. Neutrons, in contrast, are considered composite particles which, from the *de Broglie* relationship, can exhibit wave-like behaviours with λ_{neutron} ranging from ~1 to 50 Å for the type of scattering experiment discussed here.

If we consider a typical soft matter system, say polymer chains in a solvent, we need to straddle several orders of magnitude in length scale to explore all possible macromolecular phenomena. For example, to investigate bond orientations and/or side group motions we would need a probe that has a wavelength of a few Å. To extract information about the polymer chains' radius of gyration we would need perhaps λ_{probe} = ~10–200 Å and to study, say, the diameter of aggregates then λ_{probe} may need to range from 1000–10 000 Å.

So, a combination of light, X-ray and neutron scattering would be advantageous if one wishes to extract the widest possible spatial information. In fact, since neutrons and X-rays have comparable wavelengths one might argue that perhaps we don't need to use both. However, as we shall discuss, there are several important advantages of using neutrons. While X-rays (and light) interact electromagnetically with the electron cloud, neutrons are scattered by short-range nuclear forces (Figure 2.2). This fundamental difference leads to key rewards when using the neutron as a probe of matter:

1. The nucleus has a diameter of the order of a femtometre, 10^{-15} m; ~10^5 times smaller than λ_n. As a result, to the neutron, the nucleus appears as a single point of negligible size in space. The intensity of the scattered neutron wave therefore shows no angular dependence (*i.e.* no form factor) and corrections to compensate for non-isotropic scatter need not be applied. In contrast, the electron shell has a diameter comparable to $\lambda_{\text{X-ray}}$ (*i.e.* 0.1–10 × $\lambda_{\text{X-ray}}$). The scattered X-ray intensity therefore shows a marked angular variation that requires modification.

2. Neutrons, being neutral, are unimpeded by the electron cloud and are thus able to penetrate deep into matter. As a result, complex 'sample environment' apparatus (*i.e.* furnaces or high-pressure systems which require thick-walled containment vessels) can be used without significant neutron flux attenuation.

3. The neutron can be viewed as a non-destructive probe. By virtue of wave-particle duality, a 1.5 Å neutron has an associated energy of E_n = 0.036 electron-volts (eV). In contrast, a 1.5 Å X-ray has an incident energy, $E_{\text{X-ray}}$ = 8265 eV (*i.e.* ~200 000 × E_n). If we consider that a typical bond energy is of the order of a few electron-volts

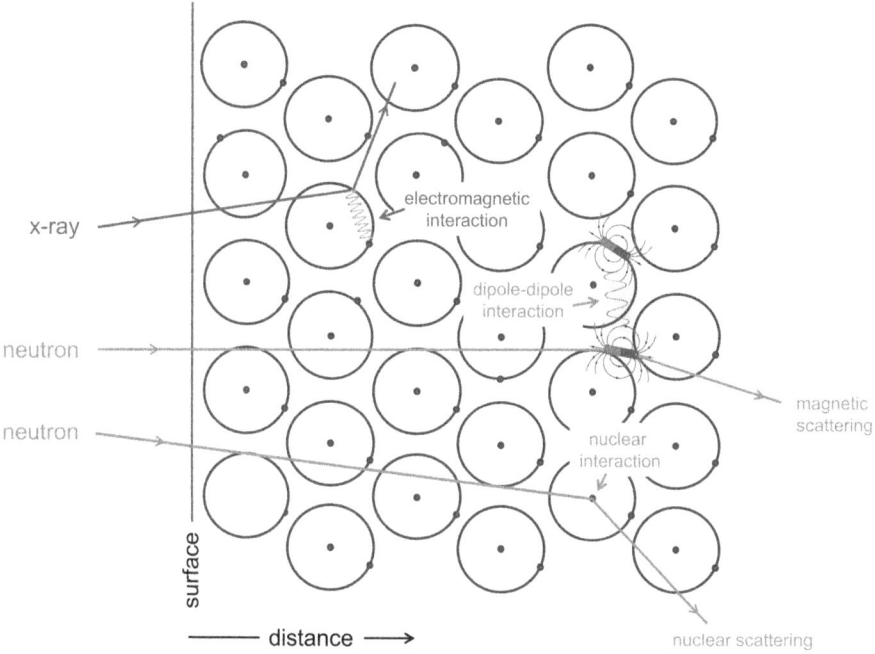

Figure 2.2 The scattering mechanisms that drive X-rays and neutron (nuclear and magnetic) scattering.[2]

(*e.g.* the bond-dissociation energy for a single carbon–hydrogen bond in ethane is ~4 eV/bond), then X-rays can quickly cause molecular degradation. In addition, the fact that a system is weakly perturbed by the neutron greatly facilitates theoretical interpretation of neutron scattering data.

4. Finally, the intensity of the scattered neutron wave, which is governed by the neutron–nucleus interaction, is modulated 'randomly' across the periodic table. In contrast, the intensity of a scattered X-ray varies linearly with number of orbital electrons; scattering from the hydrogen atom (^{1}H) being negligible compared to the strong signal arising from, say, gold (^{79}Au). This difference has important consequences which will be highlighted later.

Note: The neutron, with its own magnetic moment of μ_n = 5.051 × 10^{-27} J T^{-1}, can couple (*via* a dipole–dipole interaction) to, and be scattered by, unpaired elections. The neutron as a probe is therefore of fundamental importance for the study of magnetic materials and associated phenomena.

Of course, no one probe is supremely unique. Nonetheless, the aforementioned advantages counterbalance the following limitations:

1. Neutron sources lack 'brilliance'. Flux at the sample position on a neutron instrument can be orders of magnitude less than that observed at an X-ray facility. Consequently, the amount of sample required for a neutron experiment is generally greater and/or measurement times are considerably longer.
2. While useful for spectrometer shielding, and for minimising potentially problematic 'background' signal, some elements in the periodic table absorb, rather than scatter, neutrons *e.g.* cadmium, boron, gadolinium.
3. It is tricky to focus a neutron beam. Neutron sample sizes are therefore typically of the order of a few square centimetres (cm^2); a challenging dimension for materials of limited availability and/or high cost. Collimation can be used to match the neutron beam profile to the sample size but to the detriment of incident neutron flux. In contrast, for X-ray measurements, samples can be of the order of a few square millimetres in size (mm^2).

As a visit to any neutron or bright X-ray source demonstrates, implementing both experimental techniques requires, at present, complex and sizeable infrastructure housed on dedicated research parks. Indeed, the cost of such 'central facilities' is shared largely between nations. However, it must also be noted that the installation of both types of facility on a single campus does promote important and beneficial synergies between the two methods.

2.3   The Scattering Process

Now that we have considered the pros and cons of the neutron as a probe of matter let's dissect the scattering process. Consider the simple neutron spectrometer and scattering experiment below. For the purpose of this scattering illustration, the instrument described is a simplified direct geometry neutron spectrometer (see Section 3.2.2) operating in time-of-flight (ToF) mode, *i.e.* a class of instrument that measures neutron–nucleus interactions in *terms of energy* by timing how long it takes for each neutron to travel from source to sample, from sample to detector. Here, the neutrons first travel a distance, L_1, interact in some manner with the material and are scattered towards one of '*N*' detectors. Each detector is set at a different scattering angle, θ, relative to the incident beam direction but placed at the same radial distance, L_2, from the sample position. For clarity, only Detector 1 and Detector *N* are shown in Figure 2.3.

We will assume that our neutron instrument is ultimately sensitive (*i.e.* unhindered by the source and/or neutron-optic related resolution constraints discussed later) and that, for this illustration, we are interested in the dynamical behaviour of the sample.

What signal might we expect to observe in, say, Detector 1?

Let's filter the neutron energy spectrum produced by the neutron source such that a single neutron, with a well-defined energy, E_i, enters the instrument. If we do this, and define the time at which this neutron passes the filter as $t = 0$, then we can record how long the particle takes to travel from source to Detector 1, *i.e.* $t_{arrival}$, following its interaction with the sample. After collision with a nucleus, the neutron will leave the sample with a final energy, E_f. Once the neutron has reached

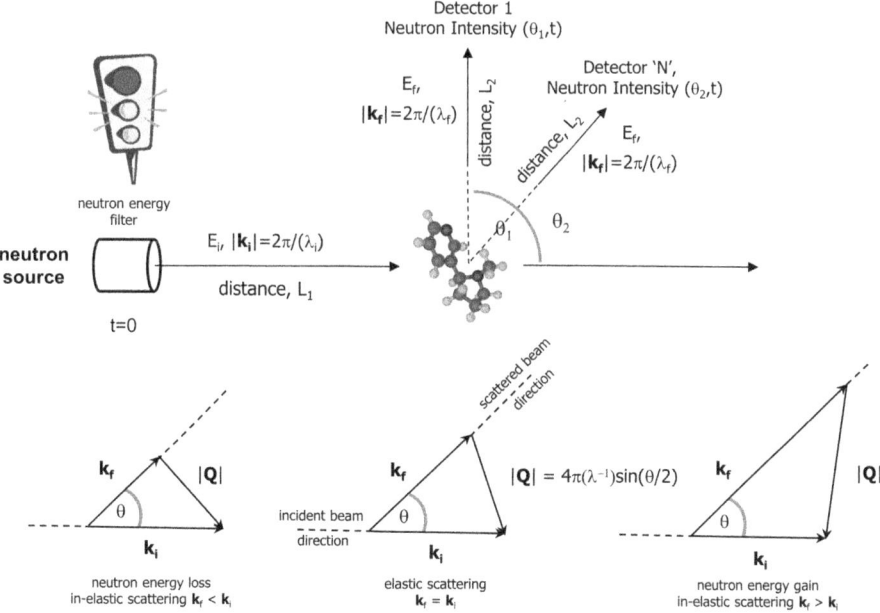

Figure 2.3 A simple neutron scattering setup. As explained in Section 2.4, the neutron's incident and scattered wave vectors are $\boldsymbol{k}_{i,f}$ (whose magnitudes are given by $|\boldsymbol{k}_{i,f}| = k_{i,f} = 2\pi/\lambda_{i,f}$) with the resulting momentum transfer vector, $\boldsymbol{Q}$, being dependent upon the energy gained or lost during the neutron–nucleus collision.

the detector, and been time-stamped by the instrument electronics, we then repeat the process.

The energy and momentum of the incident neutron are given by,

$$E_i = \frac{1}{2}m_n v_i{}^2 \text{ and } p_i = m_n v_i = \frac{h}{\lambda_i} \tag{2.1}$$

where the *de Broglie* wavelength, λ_i (the wavelength associated with a particle of rest mass, m_n), is related to its momentum, p, through the Planck constant, h (6.626×10^{-34} J s). m_n is the mass of the neutron ($m_n = 1.6749 \times 10^{-27}$ kg) and v its velocity. The distances from source to sample, L_1, and from sample to detector, L_2, are accurately known since neutron instruments are surveyed precisely when constructed. Consequently, the time, t_i, for a neutron of energy, E_i, to reach the sample position is simply given by,

$$E_i = \frac{1}{2}m_n v_i{}^2 = \frac{1}{2}m_n \left(\frac{L_1}{t_i}\right)^2 \tag{2.2}$$

A similar expression can be written for t_f, *i.e.* the time the scattered neutron of energy, E_f, takes to travel a distance, L_2, to Detector 1. We can therefore relate the amount of energy transferred between the neutron and nucleus during collision as,

$$\Delta E = \Delta\hbar\omega = E_i - E_f \tag{2.3}$$

which, by noting that $t_{\text{arrival}} = t_i + t_f$, equates to,

$$\Delta E = \Delta\hbar\omega = E_i - E_f = \frac{1}{2}m_n \left[\left(\frac{L_1}{t_i}\right)^2 - \left(\frac{L_2}{t_{\text{arrival}} - t_i}\right)^2\right] \tag{2.4}$$

Here, $\hbar$ ($= h/2\pi$) is the reduced Planck constant ($1.054571817 \times 10^{-34}$ J s).

Note: You should be aware that in the scientific literature, energy transfer may be denoted ΔE, $\Delta\hbar\omega$ or simply $\hbar\omega$. In this work, for consistency, and unless stated otherwise, we will use $\Delta\hbar\omega$.

Let's first assume that the sample under investigation is 'immobile', *i.e.* there are no atomic or molecular processes occurring on the pico to nanosecond timescale measurable using the quasi-elastic scattering method.

In this case, to conserve energy, all neutron–nucleus interactions must be *elastic* and $E_i = E_f$. We say that the *energy transferred* between neutron and nucleus *is zero* or $\Delta\hbar\omega = 0$. A plot of detected neutron intensity in Detector 1 would simply be a delta function, known as the *elastic line*, anchored at a single well-defined time of arrival. In practice, and depending on the neutron instrument class being used, we use eqn (2.4) to convert time to *energy transfer* and plot the detected intensity as a function of $\Delta\hbar\omega$ (Figure 2.4).

> **Note:** As will be discussed later (Chapter 3), the neutron spin echo class of instrument also considers changes in energy between the incident and scattered neutron. However, here, energy exchange is determined by tracking the loss of neutron beam spin polarisation during collision, a response which is followed in terms of Fourier time, t_F.

What happens, however, if the sample is 'moving' on a timescale that matches the experimental time window of the neutron instrument?

If the kinetic energy of the mobile entity in our sample is comparable to that of the incident neutron, E_i, then energy may be transferred to, or from, the incident neutron. If so, E_f may no longer equal E_i *i.e.* $\Delta\hbar\omega \neq 0$ since $E_i \neq E_f$. Here, the neutron may 'speed up' or 'slow down' during the neutron–nucleus interaction and thus will arrive at Detector 1 before (*i.e.* neutron velocity increases), or after (*i.e.* neutron velocity decreases), we expect it to had the scattering event been *elastic* (*i.e.* neutron velocity remains constant).

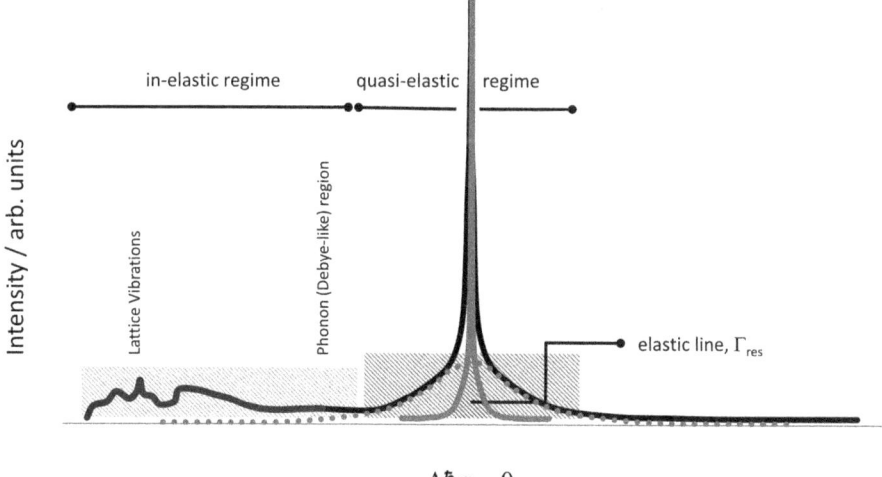

$$\Delta\hbar\omega = 0$$

Figure 2.4 Illustrative spectral contributions to $\Delta\hbar\omega$. Elastic-, quasi- and in-elastic scattering regimes observed in a single detector. The intrinsic broadening, Γ_{res}, of the elastic scattering line ($\Delta\hbar\omega = 0$) defines the sensitivity, or resolution, of a neutron instrument and limits the upper observable experimental time window, t_{res}.

As more and more neutron–nucleus collisions occur, and more and more energy transfer processes take place and more and more neutrons reach the detector, a distribution of arrival times will ensue. In this case, a plot of detected neutron counts *vs.* energy transfer in Detector 1 will resemble the peak (intensity, I, width, Γ, centred about $\Delta\hbar\omega$ = 0) we considered earlier (Figure 2.1). A similar response will be seen in the other detectors encircling the sample position; the observed width and intensity varying, however, from detector to detector in a fashion dependent upon the fundamental and unique behaviour of the sample at the nanoscale.

In reality, just because $E_i \neq E_f$ it doesn't always follow that we are probing quasi-elastic scattering phenomena. Depending upon the type of excitation probed, and the amount of energy transferred upon collision, we can categorise non-elastic scattering processes as either *quasi-* (typically $\Delta\hbar\omega$ < 2 meV) or *in-elastic* (aka INS, typically $\Delta\hbar\omega$ > 2 meV). The latter are usually associated with scattering phenomena of distinct frequency, which impart energy transfer events that are usually well separated from the elastic line (Figure 2.4).

Furthermore, no neutron instrument exhibits such sensitivity that it can measure to infinite time as we have assumed so far. As Figure 2.4 illustrates, elastically scattered events don't in fact result in a delta-function at $\Delta\hbar\omega$ = 0, but instead a peak of intrinsic width, Γ_{res}. While the construction of any neutron instrument attempts to limit this width, it is governed by unavoidable design limitations. Knowing Γ_{res} is fundamental since it defines the resolution of an instrument and provides a measure of the upper experimental observation time of a spectrometer, as will be discussed in Chapter 3.

2.4   Momentum Transfer, **Q**, and **Q**-ω Space

In the previous section we considered a simple quasi-elastic scattering scenario. We saw that the detected neutron signal resembles a peak (in energy) of height, I, and of width, Γ. The width and height of this peak may vary from one detector to the next (*i.e.* as a function of scattering angle, θ and ω) in a manner dependent upon, and unique to, the material under investigation. In practice, however, we want to relate the detected signal to quantities that define how the neutrons interact with the nuclei. Instead of analysing intensity as a function of scattering angle, θ, therefore, we relate θ to the change in neutron momentum during collision. If we define the neutron's incident and scattered wave vectors as k_i and k_f (whose magnitudes are given by $|k_{i,f}| = k_{i,f} = 2\pi/\lambda_{i,f}$) then,

$$Q = k_i - k_f \tag{2.5}$$

where Q = the momentum transfer vector.

Using simple vector operations on k_i, k_f and θ, we can construct scattering triangles, and use the cosine rule, to determine the magnitude of the momentum transfer vector, $|Q|$, for any combination of E_i and E_f such that,

$$|Q|^2 = (2\pi/\lambda_i)^2 + (2\pi/\lambda_f)^2 - 2(2\pi/\lambda_i)(2\pi/\lambda_f)\cos(\theta) \tag{2.6}$$

For elastic scattering this reduces to,

$$|Q|_{k_i = k_f} = \frac{4\pi}{\lambda_i}\sin\left(\frac{\theta}{2}\right) \tag{2.7}$$

since $E_i = E_f$ and $k_i = k_f$. It should be noted that when referring to the $|Q|$ value associated with a specific neutron detector, the scalar value quoted is always that related to the elastic scattering condition (eqn (2.7)).

Note: Momentum transfer vectors are quoted in units of inverse Ångström (Å^{-1}) and from here-on-in the magnitude of the momentum transfer vector, $|Q|$, will be denoted, Q. It is also worth noting that, in terms of wave vectors, the energy transferred during collision is given by,

$$\Delta\hbar\omega = \frac{\hbar^2}{2m_n}\left(k_i^2 - k_f^2\right) \tag{2.8}$$

We can use eqn (2.6) to determine Q for any combination of E_i, E_f and θ, and thus construct a region in parameter space, aka Q–ω space, to explore during our experiment. The relevance of, and the relationship between, Q–ω space and the physical properties of a material will become apparent. However, one should appreciate from eqn (2.6) that the Q and $\Delta\hbar\omega$ ranges accessible on any neutron instrument are limited by an instrument's operating characteristics. For example, consider our simple scattering scenario. A neutron with a single incident energy, E_i, can gain (from the sample) energy during the scattering process. Of course, it can also lose energy, but clearly no more than E_i otherwise the neutron's velocity will tend to zero. Figure 2.5 illustrates this by computing, using eqn (2.6), the Q–ω parameter space afforded by a direct-geometry instrument with eight neutron detectors subtending scattering angles, θ, from $10°$ to $150°$ ($E_i = 1.88$ meV, $\lambda_i = 6.6$ Å).

The figure also shows, for contrast, the Q–ω space accessible (broken black lines) using a different type of quasi-elastic neutron instrument; the indirect geometry spectrometer. As will be discussed in Chapter 3, this instrument operates by defining a single final energy, E_f. Clearly, as the different Q–ω detector trajectories show, no one detector measures a common momentum transfer, Q, value as a function of energy transfer; only constant θ. Good practice therefore dictates that, no matter how small the variation in momentum transfer in a single detector, to construct a data set at constant Q one should sample Q–ω space during data reduction (Chapter 6) such that data

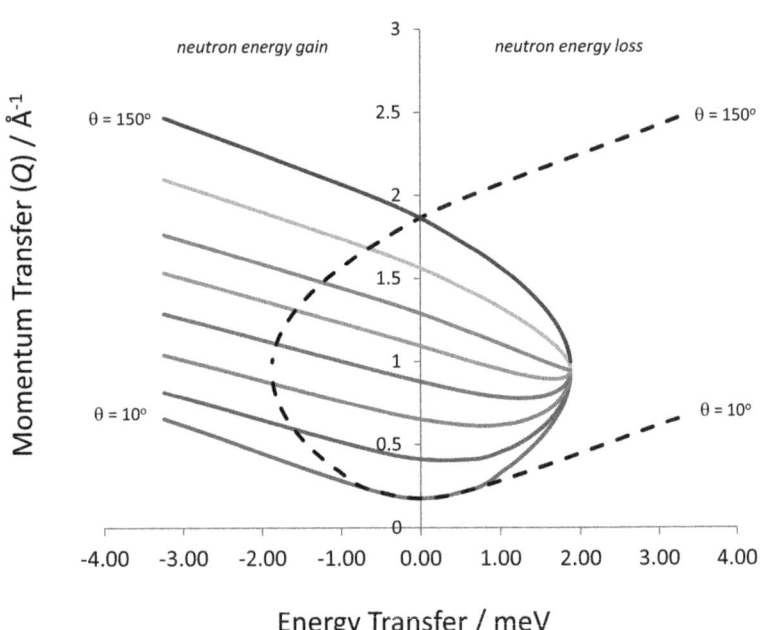

Figure 2.5 Illustrative example of Q–ω parameter space. Trajectories calculated using eqn (2.6), with $E_i = 1.88$ meV and a final scattered neutron energy range of: $E_f = 5.1$ to 0.1 meV (*i.e.* $\lambda_f = 4$–30 Å). For contrast, the Q–ω space accessible on an indirect geometry instrument that operates by defining, and detecting, a single final neutron energy ($E_f = 1.88$ meV, $\lambda_f \sim 6.6$ Å) is delineated by the broken lines.

is collated from different detectors. Of course, for those experiments that only need monitor the number of neutrons scattered elastically then such correction is not necessary. Examples of such elastic scattering measurements are given in Chapter 7.

2.5 Momentum Transfer, Q, and Length Scale Probed, l

Most QENS instruments employ an array of fixed position neutron detectors that subtend different scattering angles relative to the incident beam direction and collect data simultaneously during a measurement. A question therefore arises. If QENS spectra are collected concurrently as a function of scattering angle then,

What length scale does each detector 'probe'?

When a neutron is scattered it is 'encoded' with information about motion, and possibly structure, over a specific spatial range. The length scale over which a detected neutron has 'sampled' a material depends upon θ. If we recall that scattering angle is related to momentum transfer, Q, by,

$$Q_{k_i = k_f} = \frac{4\pi}{\lambda_i}\sin\left(\frac{\theta}{2}\right)$$

(2.9)

then, *via* its association to Bragg's law, $\lambda = 2d\sin(\theta/2)$, we find that the effective length scale probed (l) is inversely proportional to scattering angle such that $l = 2\pi/Q$. As illustrated in Figure 2.6, neutrons scattered towards detectors subtending large scattering angles (*i.e.* high Q values) carry with them information about the sample over distances smaller than those neutrons scattered towards detectors subtending shallower scattering angles (*i.e.* low Q values). Consequently, a detector array collects data over a varied spatial range in a single measurement with the actual range accessible depending upon an instrument's specific setup.

2.6 The Scattering Function, S(Q,ω)

As mentioned, we usually don't have just one detector collecting data during a QENS experiment but many running concurrently; the intensity and/or breath of the recorded data sets varying as a function of Q.

Slower motion(s)   ⬅⸺⸺➡   Faster motion(s)

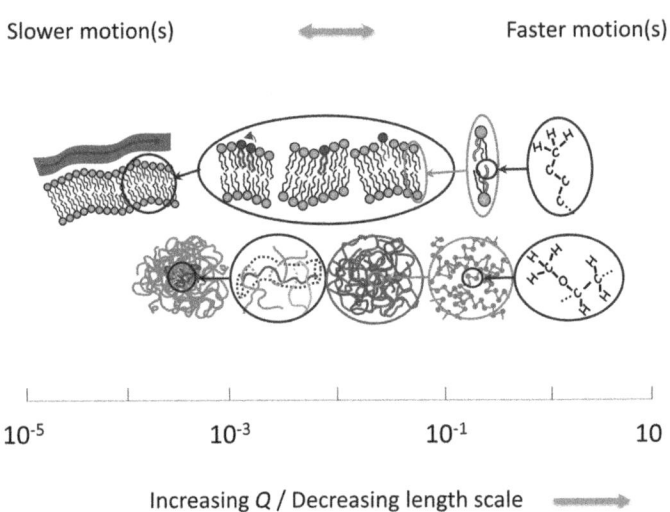

| 10⁻⁵ | 10⁻³ | 10⁻¹ | 10 |

10^{-5} 10^{-3} 10^{-1} 10

Increasing Q / Decreasing length scale   ⸺➡

Figure 2.6   Illustrative variation of length and timescale information as a function of momentum transfer; from whole molecules (low Q) to side groups (high Q).[3]

The number of neutrons scattered towards, and detected by, each detector, however, is probabilistic *i.e.* dependent upon the likelihood that a neutron of incident energy E_i is scattered by a nucleus about a direction Ω (*i.e.* a direction that points towards the detector), into a solid angle element $d\Omega$ (*i.e.* an element that matches the detector's active area) and with a change of energy $\Delta\hbar\omega = E_i - E_f = \hbar\omega - \hbar(\omega + d\omega)$. This simple statement defines the *double differential scattering cross-section*, $d^2\sigma/(d\omega d\Omega)$.

Ultimately, you want to extract 'something' from your raw neutron scattering data that contains information solely about the sample. This 'something' is the scattering function, $S(Q,\omega)$, and it so happens that the probabilistic reasoning that defines the *double differential scattering cross-section* is proportional to $S(Q,\omega)$,

$$\frac{d^2\sigma}{d\Omega d\omega} \propto \frac{k_f}{k_i}\sigma_T S(Q,\omega) \tag{2.10}$$

where σ_T is the total scattering cross section of all bound (*i.e.* no recoil effects) scattering centres which, as we shall see, is a property of the neutron–nucleus interaction. More commonly known as the dynamic structure factor, $S(Q,\omega)$ is a Q and ω dependent map of dynamic, and possibly associated geometric, nanoscale behaviour in the sample *alone*.

In reality, $d^2\sigma/(d\omega d\Omega)$ is a measure of the true dynamic structure factor $S(\mathbf{Q},\omega)$ convolved with the resolution function, $R(\mathbf{Q},\omega)$, of the neutron instrument used. For spectrometers measuring in terms of energy,

$$S^{\text{expt}}(\mathbf{Q},\omega) = R(\mathbf{Q},\omega) \otimes S(\mathbf{Q},\omega) \tag{2.11}$$

such that,

$$\frac{d^2\sigma}{d\Omega d\omega} \propto \frac{k_f}{k_i}\sigma_T S^{\text{expt}}(\mathbf{Q},\omega) \propto \frac{k_f}{k_i}\sigma_T R(\mathbf{Q},\omega) \otimes S(\mathbf{Q},\omega) \tag{2.12}$$

$\otimes$ is the convolution product. It is worth asking ourselves what happens if we aren't interested in analysing the energy of each detected neutron. What if we just want to count the number that arrive at a detector over a certain time period, as we do in a diffraction experiment? In this case we measure the *differential cross section*, $d\sigma/(d\Omega)$, and consequently a *static structure factor*, $S(\mathbf{Q})$, which contains information about the structure of the material over the spatial range covered by Q.

2.7   The Scattering Cross Section, σ

Before we continue, it is worth considering the meaning, and relevance, of the total atomic scattering cross section, σ_T. Knowledge and understanding of cross sections is imperative if one is to prepare for, and perform, a successful and accurate quasi-elastic scattering experiment. Indeed, information about the scattering cross section is necessary for the optimisation of sample thickness, absorption corrections and to ensure the absolute normalisation of units.

To do so, let's consider another hypothetical scattering experiment. Using the scattering geometry already described, let's assume that the sample is now a single, immobile (elastic scattering) atom. 'N' neutrons, of energy, E_i, are incident upon the atom and we record the number that reach, say, the first detector. We then change the atom type and repeat the exact same measurement. We do this multiple times until we have detected neutron intensity data for each element in the periodic table, including isotopes. If each measurement is identical, bar atom type at the sample position, will the same number of neutrons reach the detector?

The answer as you will probably have realised is *no*. The reason for this is that the probability that a neutron will pass by, or interact with, a nucleus is dependent upon its proximity to the nucleus, the Born approximation of the Fermi pseudopotential, $V(r) = (2\pi\hbar^2/m_n)\delta(r)b$, and the neutron ($s = 1/2$) and nucleus ($I$) spin state. Here, r is the distance of the neutron from the nucleus and b the bound (*i.e.* a particle has a tendency to remain localised in space upon collision) scattering length. b is a predominantly energy independent scattering parameter that shows seemingly random variation across the periodic table. The scattering length can be complex ($b = b' - ib''$) and may be either positive or negative. While the imaginary component is suggestive of neutron absorption, a positive or negative sign for the real component is indicative of the attractive or repulsive nature of the neutron–nucleus interaction. If interaction does occur, then the neutron may either be absorbed or undergo isotopic scatter; scatter being either coherent (interacting waves) or incoherent (non-interacting waves). The consequences of the two neutron spin (s) and 'n' nucleus spin (I) states on the scattering process is considered in detail in Sears.[4] However, some key properties are worth summarising here. First, b is defined as,

$$b = b_c + \frac{2b_i}{\sqrt{I(I+1)}}s.I \tag{2.13}$$

where b_c and b_i are the bound coherent and incoherent scattering lengths. Clearly if the nuclear spin is non zero, then b is in general spin dependent. If $I = 0$ then there is no source of incoherent scatter, *i.e.* no bound incoherent scattering length.

Second, it is more common during an experiment to consider the neutron–nucleus scattering process in terms of an effective area of interaction, *i.e.* scattering and absorption cross sections (σ_s and σ_a). Scattering cross sections have the dimension *barns* ($\times 10^{-24}$ cm^2) and enable the relative numbers of scattered and absorbed neutrons to be determined. For example, if I_0 neutrons per cm^2 per second hit a sample then the number of neutrons scattered or adsorbed can be written $I_s = \sigma_s I_0$ and $I_a = \sigma_a I_0$, respectively. Considering a single nucleus, the scattering cross section is defined as,

$$\sigma_s = 4\pi\langle|b|^2\rangle \tag{2.14}$$

and,

$$\sigma_a = \frac{4\pi}{k_i} <b''>$$

(2.15)

where k_i is the incident neutron wave vector and the brackets denote the statistical average of b over all nuclear spins.

For most nuclei, the scattering lengths and scattering cross sections are independent of the incident neutron wave vector, k_i, in the thermal neutron region. In contrast, the absorption cross sections for incident neutron energies less than a few milli-electron-volts are inversely proportional to k_i (the so-called '1/v law'). σ_a is usually tabulated for neutrons of incident wavelength, $\lambda_i = 1.8$ Å. Scattering lengths and scattering and absorption cross sections are listed in full, as both a function of atom and isotope, in Sears.[4]

In practice, for un-polarised neutrons and/or nuclei, the scattering cross section is usually denoted, σ_T, which is composed of both bound coherent σ_c ($= 4\pi|b_c|^2$) and incoherent σ_i ($= 4\pi|b_i|^2$) cross sections due to the likely occurrence of both b_c and b_i scattering lengths,

$$\sigma_T = \sigma_c + \sigma_i$$

(2.16)

In reality, however, no sample is made up of a single nuclide. Even a mono-atomic material is most likely comprised of a mixture of naturally abundant isotopes (*i.e.* differing spin states) and thus a range of scattering lengths and, as eqn (2.13) suggests, additional sources of scattered wave incoherence. Indeed, the situation becomes more complex for multi-atom systems where there will be n different types of atom, each exhibiting its own isotope mix. As a result, an average scattering length, $\langle b_n \rangle$, which takes into account both a material's spin and isotope distribution, must be considered. If $\langle b_n \rangle$ represents the mean coherent scattering length, b_n^{coh}, then the root mean square of b_n from $\langle b_n \rangle$ is known as the incoherent scattering length; $b_n^{inc} = [\langle b_n^2 \rangle - \langle b_n \rangle^2]^{1/2}$. Clearly, therefore,

$$\sigma_T = 4\pi\langle b_n \rangle^2 + 4\pi[\langle b_n^2 \rangle - \langle b_n \rangle^2]$$

(2.17)

and consequently, the double differential scattering cross section is, in practice, made up of both coherent (*i.e.* interacting scattered neutron waves) and incoherent (*i.e.* non-interacting scattered neutron waves) dynamic structure factors,

$$\frac{d^2\sigma}{d\Omega d\omega} \propto \frac{k_f}{k_i}[\sigma_c S_c(\mathbf{Q},\omega) + \sigma_i S_i(\mathbf{Q},\omega)]$$

(2.18)

Variation of σ_c, σ_i and σ_a across the periodic table is illustrated in Figure 2.7. To simplify the representation, data is presented for (i) those elements (plus specific isotopes) commonly found in samples used for quasi-elastic scattering experiments and (ii) those elements (plus specific isotopes) that are used for neutron shielding and sample containment apparatus. Regarding the latter, while clearly needing to be machine-able, cost effective and available in sizeable quantity, one should be able to appreciate from σ_c, σ_i and/ or σ_a why specific materials are chosen for a particular purpose on a neutron instrument. As an example, the sizeable σ_a of Boron (originating from its 20% [10]B abundance) and Gadolinium (originating from its 15% [157]Gd abundance) make these materials ideal for shielding.

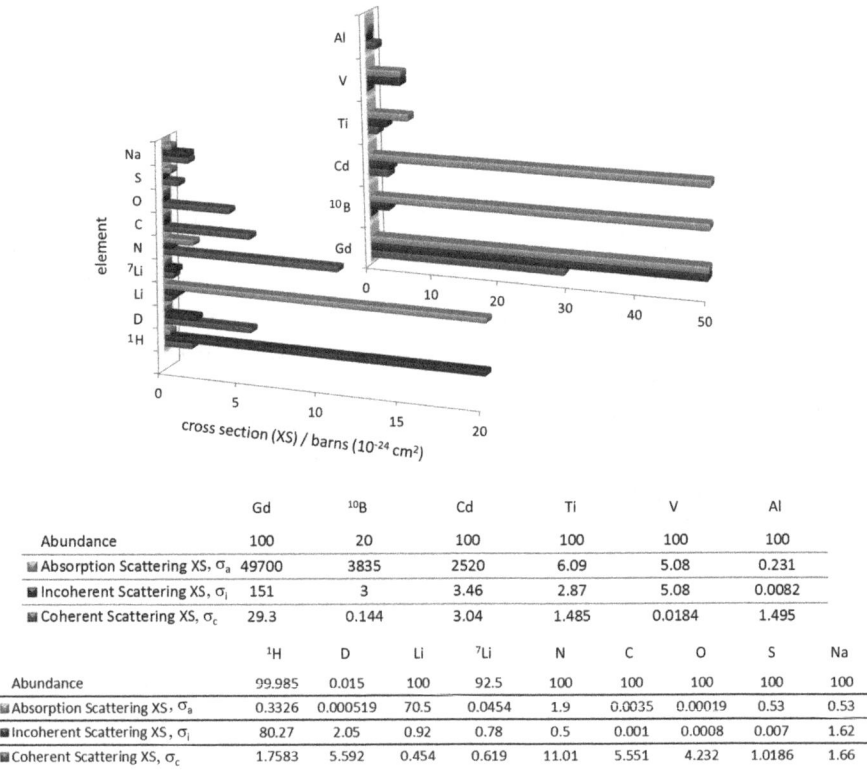

	Gd	[10]B	Cd	Ti	V	Al
Abundance	100	20	100	100	100	100
▨ Absorption Scattering XS, σ_a	49700	3835	2520	6.09	5.08	0.231
▪ Incoherent Scattering XS, σ_i	151	3	3.46	2.87	5.08	0.0082
▨ Coherent Scattering XS, σ_c	29.3	0.144	3.04	1.485	0.0184	1.495

	[1]H	D	Li	[7]Li	N	C	O	S	Na
Abundance	99.985	0.015	100	92.5	100	100	100	100	100
▨ Absorption Scattering XS, σ_a	0.3326	0.000519	70.5	0.0454	1.9	0.0035	0.00019	0.53	0.53
▪ Incoherent Scattering XS, σ_i	80.27	2.05	0.92	0.78	0.5	0.001	0.0008	0.007	1.62
▨ Coherent Scattering XS, σ_c	1.7583	5.592	0.454	0.619	11.01	5.551	4.232	1.0186	1.66

Figure 2.7 Representative variation of σ_c, σ_i and σ_a. The illustrations are truncated at 50 and 20 barns for clarity. Data taken from Sears.[4]

2.8 Interpreting $S(Q,\omega)$

The nature and form of the coherent and incoherent neutron scattering terms included in the double differential equation above (eqn (2.18)), *i.e.* $S_c(Q,\omega)$ and $S_i(Q,\omega)$, were considered by Léon Van Hove *via* analysis of the energy independent static structure factor, $S(Q)$, in 1954.[5]

Van Hove was able to correlate $S_i(Q,\omega)$ and $S_c(Q,\omega)$ with the positions of nuclei, either collectively or individually, after a given time, t, *via* self and collective intermediate scattering functions, $I_s(Q,t)$ and $I_c(Q,t)$,

$$S_c(\mathbf{Q},\omega) = \frac{1}{2\pi} \int_{-\infty}^{+\infty} I_c(\mathbf{Q},t) \exp^{(-i\omega t)} dt \tag{2.19}$$

$$S_i(\mathbf{Q},\omega) = \frac{1}{2\pi} \int_{-\infty}^{+\infty} I_s(\mathbf{Q},t) \exp^{(-i\omega t)} dt \tag{2.20}$$

which themselves are the time and space Fourier transform of the pair correlation and self-autocorrelation functions, $G(\mathbf{r},t)$ and $G_{self}(\mathbf{r},t)$,[6]

$$S_c(\mathbf{Q},\omega) = \frac{1}{2\pi} \int_{-\infty}^{+\infty} \int_{-\infty}^{+\infty} G(\mathbf{r},t) \exp^{-i(\mathbf{Qr}-\omega t)} d\mathbf{r} dt \tag{2.21}$$

and,

$$S_i(\mathbf{Q},\omega) = \frac{1}{2\pi} \int_{-\infty}^{+\infty} \int_{-\infty}^{+\infty} G_{self}(\mathbf{r},t) \exp^{-i(\mathbf{Qr}-\omega t)} d\mathbf{r} dt \tag{2.22}$$

It should be noted that the self-intermediate scattering function is often expressed as $I_s(Q,t)$ or $I_{self}(Q,t)$ in the scientific literature. Since the self-part is the essence of incoherent scattering it may also be seen denoted $I_i(Q,t)$.

Derivation of the above dynamic structure factors is given in full in Van Hove[5] and considered within the limits of classical mechanics. However, for the purpose of this work, interpretation of $S_i(Q,\omega)$ and $S_c(Q,\omega)$ can be simply summarised as follows:

The ***incoherent dynamic structure factor***, $S_i(Q,\omega)$, describes correlations between the positions of the same nucleus at different times. Arising from those detected neutron waves scattered incoherently, dynamics associated with incoherent scattering are related to uncorrelated self-motion.

In contrast, the ***coherent dynamic structure factor***, $S_c(Q, \omega)$ describes correlations between the positions of different nuclei at different times. Arising from those detected neutron waves scattered coherently, dynamics associated with coherent scattering are related to different, yet correlated nuclei and thus collective-behaviours.

It is worth noting, especially for those working at low temperatures, that to obtain a true measure of the dynamic structure factor, $S(Q,\omega)$ should be corrected using the so-called detailed balance expression, $\exp(-\hbar\omega/2k_BT)$. In addition, interpretation of correlation functions in the classical limit is valid for small energy and momentum changes. For small r (large Q) and short t (large ω) quantum effects may be expected.

Hopefully, the advantage of the seemingly random variation of scattering lengths and cross section across the periodic table is becoming apparent. To highlight the benefit further, let's consider lithium. To improve power density in battery materials there is interest in ion diffusion mechanisms in transition metal oxide materials. Clearly, a QENS experiment designed to study the self diffusion of Li ions would prove futile since naturally occurring Li has a sizeable absorption cross section ($\sigma_a^{Li} = 70$ barns) and, as such, few neutrons will scatter incoherently ($\sigma_i^{Li} = 0.92$ barns) and be detected. However, an experiment in which the experimental team replaces Li with highly abundant ^{7}Li ($\sigma_a^{7Li} = 0.0454$ barns), would, while costly, allow for a much more successful ion transport study ($\sigma_i^{7Li} = 0.78$ barns).

It should also now be apparent, that during a QENS experiment it is likely that the signal measured will contain both incoherent and coherent spectral contributions weighted accordingly. Such weighting has important implications for the study of complex multi-component materials. As example, consider hydrogenous (also referred to as protonated) materials; materials such as polymers or biological molecules that are composed of predominately carbon (C), sulphur (S), oxygen (O), nitrogen (N) and hydrogen (^{1}H). Hydrogen, as we see from Figure 2.7, exhibits a sizeable incoherent scattering cross section compared to other nuclei, $\sigma_i^H = 80.27$ barns. Indeed, σ_i^H is ~40 times greater than its own coherent cross section, $\sigma_c^H = 1.7583$ barns. As a result, any signal measured from a protonated material will be dominated by the *incoherent dynamic structure factor* $S_i(Q,\omega)$ arising from scattering from ^{1}H nuclei. The information we extract from the experiment will consequently be dominated by the mobility, or self-motion, of the ^{1}H species.

However, what if you aren't interested in investigating the ensemble average of all ^{1}H atom self-motions? Perhaps instead you just want to investigate dynamics associated with a specific chemical specie such as a methyl group, CH_3? Well, the hydrogen isotope, deuterium (^{2}H), just so happens to have a much weaker incoherent cross section, $\sigma_i^D = 2.05$ barns. As a result, by replacing with deuterium (also known as 'labelling' or 'selective deuteration') those hydrogen atoms *not* associated with methyl species, the incoherent signal, and hence dynamical information, associated solely with the CH_3 chemical group can be accentuated. For increasingly complex systems, the deuteration technique is a powerful tool that allows spectral components to be segregated. In fact, so important is deuterium labelling to the neutron method that neutron facilities embrace sample deuteration support facilities. Examples include the ILL-EMBL Deuteration Facility (D-lab) at the Institut Laue-Langevin (ILL), France, the National Deuteration Facility (NDF) co-funded by the Australian Nuclear Science and Technology Organization (ANSTO) and the European Deuteration Network (DEUNET), which involves ESS (Sweden), STFC-ISIS (United Kingdom), ILL (France) and FZJ (Germany), with ANSTO (Australia) and J-PARC (Japan) as international observer members.

To conclude, experimental incoherent dynamic structure factors, $S_i(Q,\omega)$, from telechelic poly(vinyl alcohol) (PVA) swollen hydrogels[7] containing either H_2O or D_2O are compared in Figure 2.8. Clearly, in (i), H_2O diffusion between hydrogel pores dominates. Spectral information about the polymer network mobility in the swollen state is largely masked. However, as illustrated in (ii), the use of D_2O greatly reduces the spectral intensity arising from the water component. Instead, scattering contributions due to polymer chain dynamics are accentuated, and subsequently modelled, with the hydrogel material still in its swollen form.

2.9 Spectral Information Contained Within $S(Q,\omega)$

Focusing on the incoherent dynamic structure factor, let's conclude by considering the information contained in $S(Q,\omega)$

For a bulk sample, the spectroscopic contributions that make up $S_i(Q,\omega)$ can include,

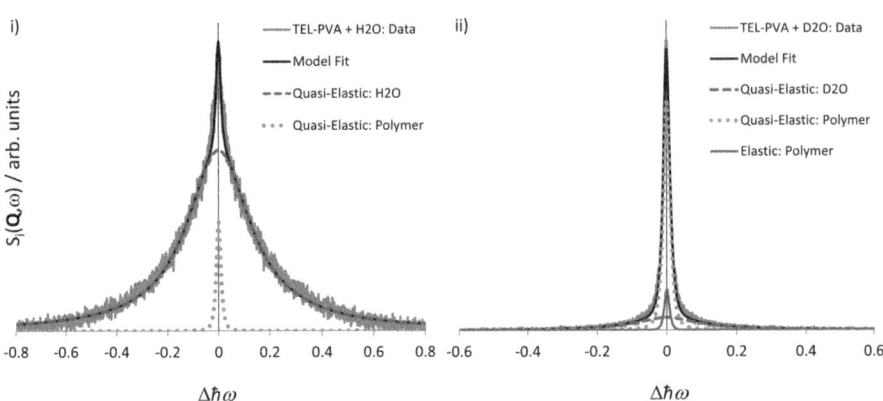

Figure 2.8 Comparison of incoherent dynamic structure factors, $S_i(\mathbf{Q},\omega)$, measured from telechelic PVA hydrogels containing (i) H_2O and D_2O (ii). $Q = 1.222$ Å^{-1} and $T = 303$ K.[7]

- lattice modes, $S_i^L(\mathbf{Q},\omega)$
- intra-molecular vibrations $S_i^V(\mathbf{Q},\omega)$
- molecular reorientations, $S_i^R(\mathbf{Q},\omega)$

For a liquid, $S_i^L(\mathbf{Q},\omega)$ should be replaced by $S_i^T(\mathbf{Q},\omega)$ to include information about long-range translational diffusion. As a result, the incoherent dynamic structure factor is denoted,

$$S_i(\mathbf{Q},\omega) = S_i^L(\mathbf{Q},\omega) \otimes S_i^R(\mathbf{Q},\omega) \otimes S_i^V(\mathbf{Q},\omega) \tag{2.23}$$

For simplicity, we assume that the different motions are independent and, in the time domain, we find that terms can be multiplied such that the self-intermediate scattering function is given by,

$$I_i(\mathbf{Q},t) = I_i^L(\mathbf{Q},t) \times I_i^R(\mathbf{Q},t) \times I_i^V(\mathbf{Q},t) \tag{2.24}$$

The situation for QENS studies can be further simplified since both $S_i^L(\mathbf{Q},\omega)$ and $S_i^V(\mathbf{Q},\omega)$ have minimal influence on the quasi-elastic regime. $S_i^V(\mathbf{Q},\omega)$ manifests itself as inelastic spectral lines peaked at high energy transfers and $S_i^L(\mathbf{Q},\omega)$ as a small background contribution to the quasi-elastic regime. However, both do contribute to the scattering function in the form of internal molecular vibrations and whole-molecule translational vibrations and librations from lattice modes, a contribution to the scattering law manifesting itself in the form of a Debye–Waller factor,

$$\text{DWF} = \exp\left(-\frac{Q^2 \langle u^2(T) \rangle}{3}\right) \tag{2.25}$$

where $\langle u^2(T) \rangle$ is the mean square displacement of the atoms under the effect of internal molecular, and external lattice modes, *i.e.* $\langle u^2(T) \rangle = \langle u_V^2(T) \rangle + \langle u_L^2(T) \rangle$. Consequently, $S_i(\mathbf{Q},\omega)$ can be written,

$$S_i(\mathbf{Q},\omega) = \exp\left(-\frac{Q^2 \langle u^2(T) \rangle}{3}\right)\left(S_i^R(\mathbf{Q},\omega) + S_i^I(\mathbf{Q},\omega)\right) \tag{2.26}$$

where $S_i^I(\mathbf{Q},\omega)$ is an illustrative inelastic contribution arising from the convolution of $S_i^L(\mathbf{Q},\omega)$ and $S_i^V(\mathbf{Q},\omega)$. As discussed above, this term contributes little to the quasi-elastic regime and is, in practice, largely neglected. For a sample that exhibits long-range translational motion, then $S_i^R(\mathbf{Q},\omega)$ should be replaced by $S_i^R(\mathbf{Q},\omega) \otimes S_i^T(\mathbf{Q},\omega)$ and $\langle u^2(T) \rangle = \langle u_V^2(T) \rangle$, such that,

$$S_i(\mathbf{Q},\omega) = \exp\left(-\frac{Q^2 \langle u^2(T) \rangle}{3}\right)\left(\left(S_i^R(\mathbf{Q},\omega) \otimes S_i^T(\mathbf{Q},\omega)\right) + S_i^I(\mathbf{Q},\omega)\right) \tag{2.27}$$

Modelling, and subsequent interpretation, of the localised $S_i^R(\mathbf{Q},\omega)$, and/or translational $S_i^T(\mathbf{Q},\omega)$ incoherent structure factors is a basic goal of a QENS experiment. While the exact spectral responses depend upon the sample under investigation, and external environmental factors, there are certain characteristics that define the two scattering functions. These characteristics become apparent if one considers the effect of the self-correlation function at infinite time. Consider two atoms. The first diffuses within a space that is very large compared to the inter-atomic distance. The second is constrained to a move within a small finite volume. As $t \rightarrow \infty$ the self-correlation function for the former, $G_{\text{self}}(\mathbf{r},\infty)$, tends to zero. In contrast, for constrained movement $G_{\text{self}}(\mathbf{r},t)$ tends towards a finite, time-independent, value. If the self-correlation function is split into these time-dependent and time-independent terms, $G_{\text{self}}(\mathbf{r},t) = G_{\text{self}}(\mathbf{r},\infty) + G'_{\text{self}}(\mathbf{r},t)$, then the intermediate scattering function can be written,

$$I_i(\mathbf{Q},t) = I_i(\mathbf{Q},\infty) + I_i(\mathbf{Q},t) \tag{2.28}$$

from which, *via* the time Fourier transform,

$$S_i(\mathbf{Q},\omega) = S_i^{\text{el}}(\mathbf{Q},\infty)\delta(\omega) + S_i^{\text{qe}}(\mathbf{Q},\omega) \tag{2.29}$$

i.e. the sum of a purely elastic, $S^{el}_i(Q,\infty)\delta(\omega)$ component superimposed on a structure factor with non-vanishing broadening. Scattering from a dynamically disordered system is characterised by the absence of the elastic component since $G_{self}(\mathbf{r},\infty)$ vanishes. In contrast, the existence of an elastic contribution signals the presence of a scattering centre, such as a caged molecule, constrained in space. For both cases, the width, and Q-dependence, of $S^{qe}_i(Q,\omega)$ contains information about the characteristic time of the associated motion. However, for the latter, the relative integrated intensities of the elastic, $I^{el}(Q)$, and quasi-elastic, $I^{qe}(Q)$, components also contain information about the geometry of the constrained motion. Such information is revealed *via* the so-called elastic incoherent structure factor (EISF), $A_0(Q)$,

$$\text{EISF} = A_0(Q) = \frac{I^{el}(Q)}{I^{el}(Q) + I^{qe}(Q)} \tag{2.30}$$

The EISF is indicative of a specific type of motion localised in space. Careful extraction and analysis of the measured EISF can therefore expose the associated geometry of motion *via* comparison with theoretical predictions. For analysis performed in the time domain, $A_0(Q)$ can be evaluated by considering the Q dependence of the long-time asymptote reached by the intermediate scattering function (*i.e.* $I_s(Q,t) \to \infty$); see Chapter 8.

Ideally, the elastic and quasi-elastic spectral integrals are evaluated for $\Delta\hbar\omega = \pm\infty$. If so, for a single quasi-elastic process, the sum of the relative intensities, $A_0(Q)$ and $A_1(Q)$, should be Q independent. As a result, eqn (2.29) can be written in terms of elastic and quasi-elastic incoherent structure factors,

$$S_i(\mathbf{Q},\omega) = \exp\left(-\frac{Q^2\langle u^2(T)\rangle}{3}\right)\left(A_0(Q)\delta(\omega) + A_1 S^{qe}_i(\mathbf{Q},\omega)\right) \tag{2.31}$$

Here, the quasi-elastic component is described using a Lorentzian function, $L(Q,\omega)$. However, since no neutron instrument straddles such an infinite energy transfer window, care must be taken when evaluating the EISF experimentally.

Traditionally, QENS experiments have largely focused on self-motion *via* analysis of the incoherent dynamic structure factor (or self-intermediate scattering function). Such dominance is due to the fact that many problems encountered using the QENS method are ones in which incoherent scattering dominates due to the sizeable incoherent signal arising from the ^1H atom. Nonetheless, collective motions can, and are routinely, measured; albeit *via* much more involved interpretation and analysis of the coherent part of the correlation function. While QENS studies of collective phenomena from, for example, deuterated materials are performed on direct and/or indirect geometry spectrometers, the timescale(s) of motion expected for collective behaviour is, broadly speaking, usually better aligned with the temporal window afforded by neutron spin echo (NSE) instrumentation. If collective motions are probed, then, for the characteristic timescales accessed by the collective, $I_c(\mathbf{Q},t)$, and self- intermediate, $I_s(\mathbf{Q},t)$, scattering functions to be related, the static structure factor, $S(\mathbf{Q})$, *i.e.* the spatial (radial) distribution of atoms in a sample, should be known,[8]

$$I_c(\mathbf{Q},t) \approx I_s(\mathbf{Q},t)\left(\frac{Q}{\sqrt{S(\mathbf{Q})}},t\right) \tag{2.32}$$

One advantage of probing $I_c(\mathbf{Q},t)$ using NSE is that the intermediate scattering function is surveyed directly, as will be discussed in Chapter 3. However, of equal benefit here, and as shown in Figure 3.1, is the fact that the length scales probed by NSE align with those accessible using Small Angle Neutron Scattering (SANS). As a result, the structural $S(\mathbf{Q})$ information needed to correlate the two types of motion can be measured using SANS prior to a spin echo experiment.

References

1. V. Arrighi, S. Gagliardi, F. Ganazzoli, J. S. Higgins, G. Raffaini, J. Tanchawanich, J. Taylor and M. T. F. Telling, *Macromolecules*, 2018, **51**, 7209–7223.
2. R. Pynn, *Los Alamos Science*, Los Alamos, 1990, p. 31.
3. V. Garcia Sakai and A. Arbe, *Curr. Opin. Colloid Interface Sci.*, 2009, **14**, 381.
4. V. F. Sears, *Neutron News*, 1992, **3**, 26–37.
5. L. Van Hove, *Phys. Rev.*, 1954, **95**, 249–262.
6. R. E. Lechner, in *Neutron Scattering in Biology: Techniques and Applications*, ed. J. Fitter, T. Gutberlet and J. Katsaras, Springer, Berlin, 2006.
7. G. Paradossi, F. Cavalieri, E. Chiessi and M. T. F. Telling, *J. Phys. Chem. B*, 2003, **107**, 8363–8371.
8. D. Richter, M. Monkenbusch, A. Arbe and J. Colmenero, in *Neutron Spin Echo in Polymer Systems*, 2005, vol. 174, pp. 1–221.

3 Which Spectrometer Should I Choose?

In this chapter we will consider:

- The different instrument classes used for QENS studies.
- The currently available suite of QENS spectrometers.
- The international neutron scattering facility landscape.

3.1 Instrument Classes

Three main neutron spectrometer types, or classes, exist for the study of quasi-elastic neutron scattering. The first two, the so-called direct geometry and indirect geometry neutron instruments are, broadly speaking, design-comparable and,

- evaluate the quasi-elastic response in terms of the dynamic structure factor, $S(\mathbf{Q},\omega)$
- access length scales typically ~1–30 Å
- access a temporal range from ~1–6000 picoseconds
- operate by defining either E_i (direct geometry) or E_f (indirect geometry)
- are most commonly used to investigate local, self-motion(s).

A Practical Guide to Quasi-elastic Neutron Scattering
By Mark T. F. Telling
© Mark T. F. Telling 2020
Published by the Royal Society of Chemistry, www.rsc.org

If we were to subdivide and compare the capabilities of direct geometry and indirect geometry instruments further, then direct geometry instruments,

- afford poorer energy resolution
- access higher neutron energies and are thus more suited to probing high energy transfer inelastic phenomena as well as quasi-elastic events
- afford a wider energy transfer window
- are more attuned to measure Q–ω space in terms of neutron energy gain (see Figure 2.5).

Indirect geometry instruments are also referred to as backscattering spectrometers for reasons that will become apparent later.

The third instrument class is the neutron spin echo (NSE) spectrometer, which employs spin polarised neutron beams to track quasi-elastic scattering. Generally, this instrument type is used to:

- evaluate the quasi-elastic response in terms of the intermediate scattering function, $I(\mathbf{Q},t)$
- probe length scales up to ~ 500 Å (comparable to SANS 'structure' experiments)
- access temporal ranges approaching the micro-second (µs)
- operate using a band of spin polarised incident neutron wavelengths ($\Delta\lambda_i$)
- investigate global, collective motion(s)

Considering Figure 1.1, and by broadly subdividing the time and length scale ranges in terms of instrument type, Q–ω space straddled by current QENS instrumentation is illustrated in Figure 3.1. While not discussed here, triple axis neutron instrumentation, on which QENS measurements are rarely performed nowadays, is also included.

Clearly, no one instrument class spans such a wide spatial regime that it covers all length scales, nor exhibits such fine sensitivity as to probe all relaxation rates. As a result, probing a system's complete dynamic landscape may well require data from different neutron instruments, and possibly complementary experimental methods, to be collated; a rewarding, albeit time-consuming, approach with inherent experimental limitations.

While it is not the aim of this book to teach you how to design your own neutron spectrometer, understanding the basic operating

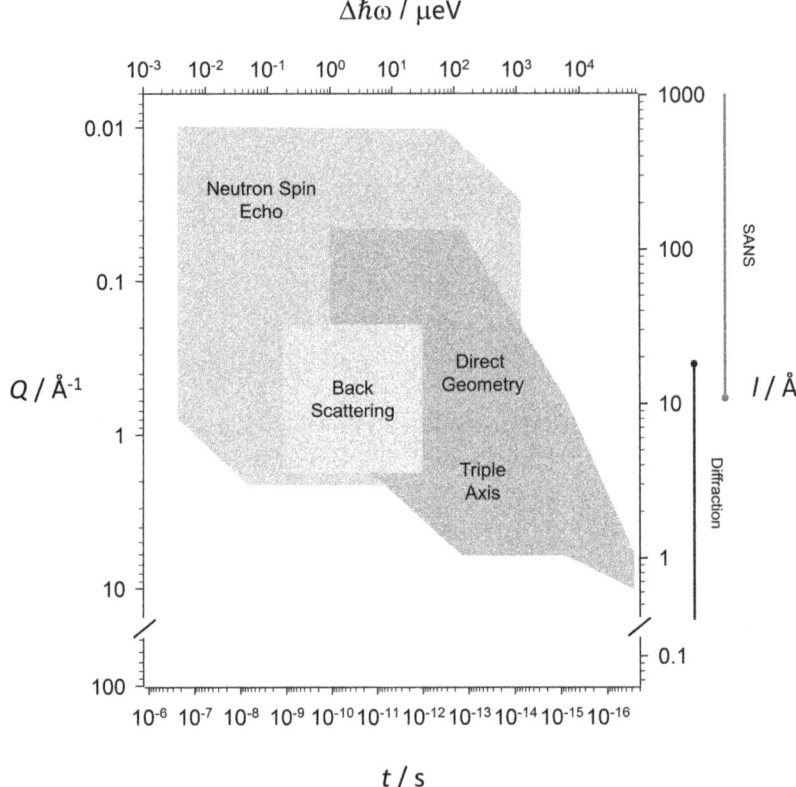

Figure 3.1 Temporal and spatial ranges broadly subdivided in terms of instrument class.

principles of these different neutron instrument types will help you decide which one is best matched to your particular scientific needs. Section 3.4 at the end of this chapter summarises those neutron science facilities which currently house QENS instruments, the spectrometers available and their routine operating characteristics. A technical review of current state-of-the-art spin-echo and backscattering spectrometers, showcasing their role in the study of nanoscale dynamics in soft and biological materials, as well as disordered magnets, is presented by Gardner et al.[1]

3.2   The Spectrometers

In Chapter 2 we considered a simple scattering experiment where the instrument described was a simplified version of a so-called 'direct geometry' chopper spectrometer running in time-of-flight (ToF) mode; quasi-elastic events, assigned in terms of the amount of energy exchanged between neutron and nucleus during collision, being determined *via* final neutron velocity discrimination. The first component of any instrument, of course, is the neutron source. Neutrons can be generated using a steady state research reactor (*i.e. via* fission of U^{235} and where neutron flux does not change appreciably with time) or a pulsed source (*i.e.* accelerator-based neutron generation *via* the spallation process where collisions between accelerated, high-energy protons and a metal target release neutrons in discrete bursts). In both cases a spectrum of neutron wavelengths emanates (see Figure 3.3), which first needs to be moderated and/or filtered is such a way that spectroscopically matched neutron energies are transported to the sample position. The manner in which filtering occurs depends upon the mode of instrument operation and is considered later.

3.2.1 Experimental Observation Time

A key consideration when deciding which neutron instrument to use is whether the experimental time window accessible matches the characteristic relaxation rate of the motion(s) to be probed. Regardless of class, the upper experimental observation time, t_{res}, afforded by any neutron instrument is dictated by its resolution.

For neutron spin echo (NSE) instruments, quasi-elastic scattering is tracked in time *via* analysis of the intermediate scattering function, $I(\mathbf{Q}, t_F)$. In the quasi-elastic limit, the Fourier time, t_F, is equivalent to real time. If not already tabulated in instrument documentation, the upper observable time limit of an NSE spectrometer can be determined directly by measuring a wholly elastic scattering calibration sample (see Section 6.3.3).

However, for direct and indirect geometry instruments, whose resolution response is defined in terms of energy and governed by a resolution function, $R(\mathbf{Q}, \omega)$, of width Γ_{res} (FWHM), such information is not immediately clear. If not tabulated, then historically two methods have been proposed for gauging t_{res}. First, and as a rule of thumb, the upper observation time can be calculated by noting its relationship to the energy resolution of a neutron spectrometer *via* the Heisenberg uncertainty principle,[2]

$$t_{res} \sim \frac{h}{2\pi(\Gamma_{res})} \tag{3.1}$$

Here the Planck constant, h, is equivalent to 4.135×10^{-15} eV s. Such association illustrates that to observe rapid motion then poor energy resolution instruments should be considered.

However, while eqn (3.1) gives a broad indication of t_{res}, experimentally this limit can extend significantly beyond the theoretical prediction, as illustrated in Figure 3.2. If not readily available, an estimation of a spectrometer's upper experimental observation time is best ascertained by measuring the incoherent dynamic structure factor, $S_i(\mathbf{Q},\omega) = R(\mathbf{Q}\omega)$, from a purely elastic scattering calibration material (aka standard). Should this data be Fourier transformed then the point at which the resulting intermediate scattering function, $I_s(\mathbf{Q},t)$, tends to zero marks t_{res}. Care must be taken when choosing where to truncate $I_s(\mathbf{Q},t)$ data since instabilities at extended time become evident. It is worth remarking that t_{res} may also be estimated by noting that 1 µeV is equivalent to 2.42×10^8 Hz such that a micro-electron volt equates to ~4.1 nanoseconds. This relationship results in a theoretical value closer to that determined experimentally. Temporal ranges, as calculated using eqn (3.1) for selected QENS instruments, are tabulated in Table 3.1. t_{res}, as determined from $I_s(\mathbf{Q},t)$ is illustrated in Figure 3.2.

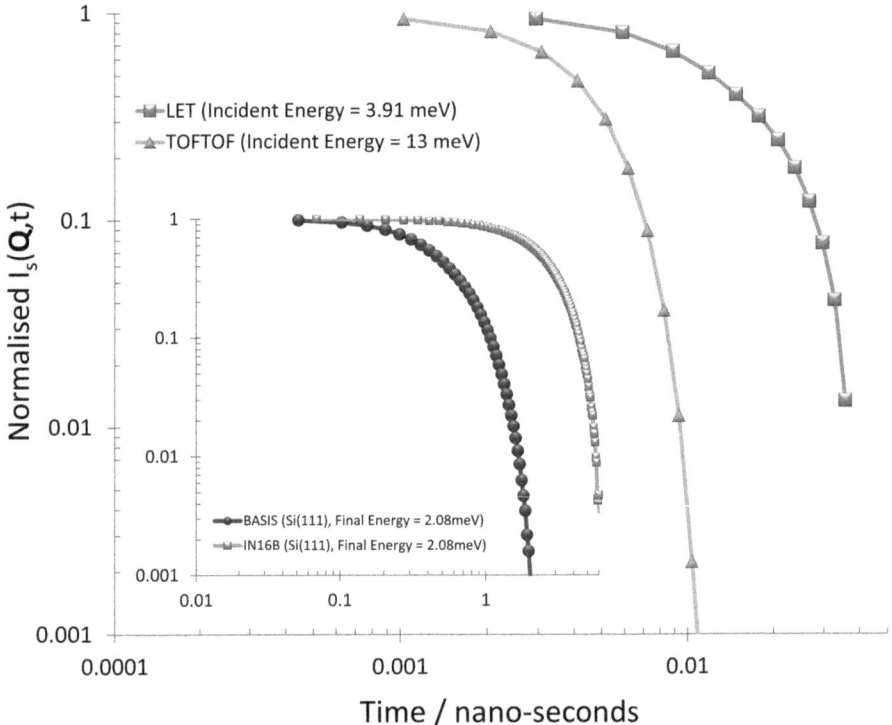

Figure 3.2 Upper experimental observation times, t_{res}, for selected neutron instruments as determined by Fourier transforming each instrument's resolution function, $S_i(\mathbf{Q},\omega)$.

Table 3.1 Upper experimental observation times, t_{res}, as determined (a) theoretically from eqn (3.1) and (b) by Fourier transforming the instruments' resolution functions, $S_i(\boldsymbol{Q},\omega)$, and noting the time at which the associated $I_s(\boldsymbol{Q},t)$ tends to zero (Figure 3.2).

Instrument name facility, country	$\varGamma_{res}$ (µeV, FWHM)	Heisenberg upper experimental observation time, t_{res}, determined using eqn (3.1)	Experiment (FFT) upper experimental observation time, t_{res}, i.e. time at which $I_s(\mathbf{Q},t)\sim0$
TOFTOF, FRM-II, Germany	~500 ($E_i = 13$ meV)	~1 ps	~10 ps
LET, ISIS, UK	~100 ($E_i = 3.9$ meV)	~7 ps	~35 ps
BASIS, SNS, USA	~3.5	~188 ps	~2000 ps
IN16B, ILL, France	~0.75	~878 ps	~6000 ps

Now we understand how to gauge the upper experimental observation limit of an instrument, let's consider the instruments themselves. We begin by considering those spectrometers operating in Q–ω space.

3.2.2 The Direct Geometry Spectrometer

A schematic of a direct geometry ToF chopper instrument is shown in Figure 3.3. The instrument consists of a primary spectrometer and a secondary spectrometer. The former transports the incident beam from source to sample while the latter carries the scattered beam from sample to detector. It is the goal of any instrument designer to optimise, and match, the spectral behaviours (*i.e.* resolution) of the two.

For direct geometry instruments, the primary spectrometer is designed to transport a highly monochromatic neutron beam to the sample position. Historically, three methods have been used. The first employs an array of velocity discriminators (*e.g.* Fermi choppers or an arrangement of neutron absorbing discs (aka disc choppers) with narrow neutron-transparent apertures) phased relative to neutron production at the source. The second method utilises the principle of Bragg scattering from an array of single crystals to define which incident neutron energy, E_i, is Bragg reflected towards the sample. The third, known as time focusing, utilises a combination of mono-chromators, neutron filters (mostly beryllium) and velocity discriminators.

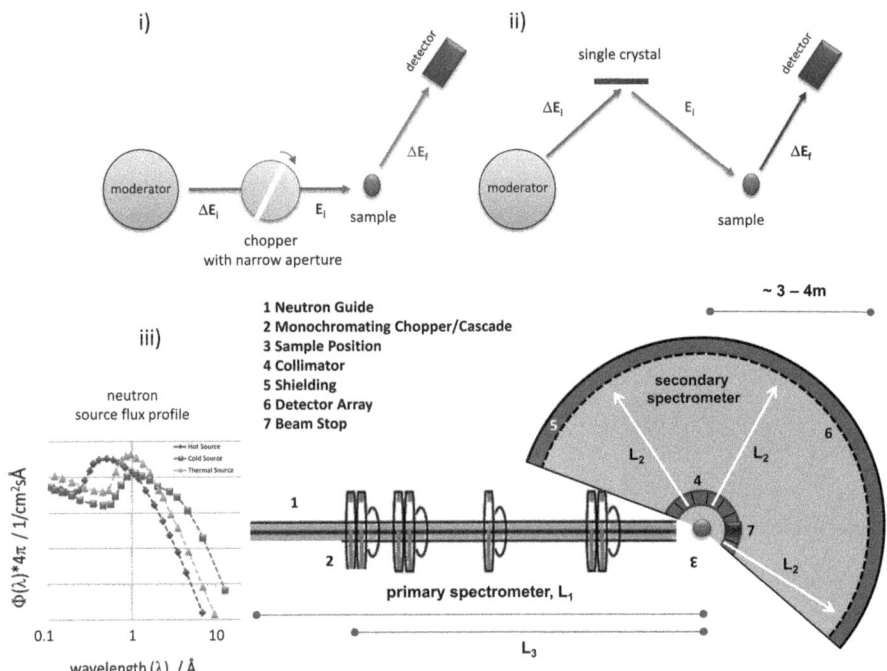

Figure 3.3   (i and ii) Direct geometry instrument schematics that illustrate different methods of filtering (aka mono-chromating) the incident beam. If constructed at a research reactor, the moderators shown would be replaced by a pulsing device to impose a timing structure. (iii) Illustrative 'bird's eye view' of a direct geometry time of a flight instrument that uses a disc chopper cascade to filter the flux distribution generated by the source. The neutron flux distributions shown are those generated by the MLZ reactor operating at 20 MW. Data taken from MLZ online resources.[3]

Direct geometry instruments in operation today run in ToF mode. Here, the arrival time of a scattered neutron at the detector is time-stamped by the detection electronics relative to a well-defined $t = 0$. At spallation institutes, ToF operation is inherent due to the pulsed nature of neutron generation. At a research reactor, however, a pulsing device (such as a disc chopper) is needed to impose a timing structure on the incident beam.

After scatter, a spectrum, or band, of neutron energies, ΔE_f, travels 'directly' to the detector bank. All spectrometer distances are accurately known and the detectors are sufficiently far away from the sample to allow the error in the ToF measurement to be minimised. By positioning a multi-detector assembly radially about the sample position a sizeable amount of temporal and spatial information can be collected concurrently.

The detectors themselves, especially on next generation or newly upgraded instruments, are long (~2 m), thin (~inch diameter) positon sensitive ^{3}He tubes; the position resolution of neutron detection along the length of the tube being the order of an inch. Such technology not only provides Q information in the horizontal scattering plane but also in the vertical. As a result, multi-dimensional scattering information can be collected from non-isotropic systems. In contrast, while positon sensitive detectors are used on indirect geometry instruments (*i.e.* BASIS, SNS[4] and DNA, J-PARC[5]) their purpose is to enhance energy resolution rather than provide addition Q information in the vertical scattering plane.

The energy resolution, or sensitivity, of a neutron instrument adds an intrinsic width to the elastic line, Γ_{res}. For a direct geometry instrument the resolution is usually defined in terms of a percentage relative to the incident neutron energy. For example, AMATERAS (J-PARC[6]) runs with an energy resolution of $\Delta E/E_i > 1\%$ for $E_i = 20$ meV (see Section 3.4). The energy resolution of a direct geometry chopper instrument arises from, and is a combination of, the uncertainties in accurately quantifying the incoming neutron velocity. On a pulsed source $\Delta E/E_i$ can be calculated using,[6,7]

$$\left(\frac{\Delta E}{E_i}\right)^2 = \left(\frac{2\Delta t_{burst}}{t_{chopper}}\right)^2 \left(1+\frac{L_1}{L_2}\right)^2 + \left(\frac{2\Delta t_{source}}{t_{chopper}}\right)^2 \left(1+\frac{L_3}{L_2}\right)^2 \quad (3.2)$$

where Δt_{burst} is the burst time of the mono-chromating chopper used to define, E_i, Δt_{source} is the time width of the pulse at the neutron source and $t_{chopper}$ is the ToF from the source to the mono-chromating chopper. The distances, L_1, L_2 and L_3 are flight paths: source to sample, sample to detector and mono-chromating chopper to sample, respectively.

The energy resolution of a direct geometry spectrometer is highly dependent on chopper speeds and the ratio of instrument lengths. While primary and secondary spectrometer lengths are optimised and fixed during the design stage, one can appreciate that by changing the chopper speed, and thus burst time, the energy resolution can be tuned, usually at the expense (gain or loss) of neutron flux at the sample positon. Variation of chopper aperture can also modify instrument characteristics.

It is worth concluding by highlighting developments in the area of multi-chopper direct geometry instruments, such as LET at ISIS,[8] in which variable resolution data sets can now be collected in a single measurement. Here, the chopper cascades on these spectrometers are phased such that a number of different, yet well defined, neutron energies are incident sequentially upon the sample. While the Q range probed, and flux available, varies as a function of E_i, instruments that operate in this repetition-rate multiplication (RRM) mode greatly extend the accessible temporal window. As an example, Figure 3.4 shows the intermediate scattering function, $I_s(\mathbf{Q},t)$, from

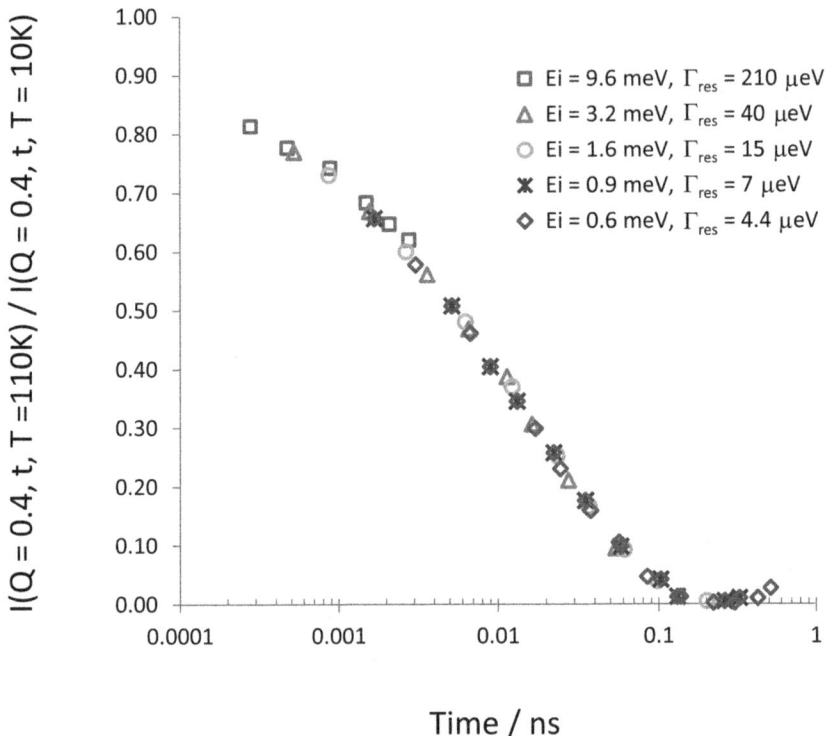

Figure 3.4 Hexadecane at 110 °C run on LET in repetition-rate multiplication (RRM) mode. $Q = 0.4 \text{ Å}^{-1}$.

liquid hexadecane measured at 110 °C on LET running in RRM mode. A measurement using five different incident energies is required to follow the relaxation.

3.2.3 The Indirect Geometry Spectrometer

In contrast to direct geometry spectrometers, indirect geometry instruments (Figure 3.5) operate by transporting a finite distribution of incident neutron energies, ΔE_i, to the sample positon and detecting a single, E_f.

Indirect geometry instruments found at research reactors (*i.e.* IN16B, ILL, France[9] or HFBS, NIST, USA[10]) generate an incident energy distribution using a mono-chromator (single crystal) placed before the sample position. The mono-chromator Bragg reflects the incident beam such that only those neutrons with an incident energy that satisfies the Bragg condition are transported towards the sample position. However, the mono-chromator itself is mounted on a mechanical Doppler unit which, when oscillating, imparts a small

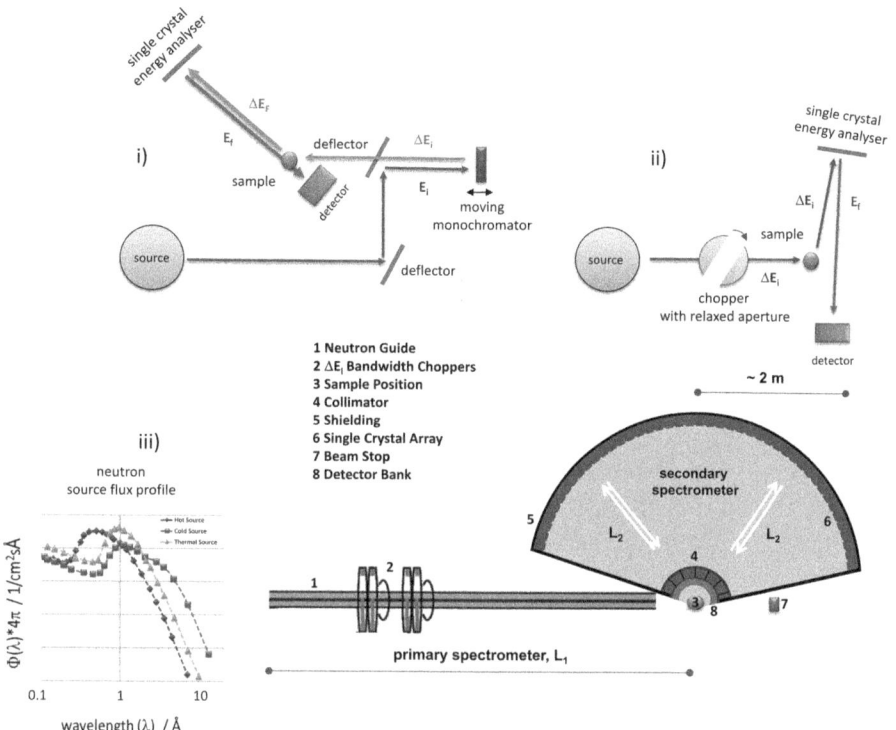

Figure 3.5 (i and ii) Indirect geometry instrument schematics showing the main methods used to define a finite incident energy distribution and, subsequently, isolate E_f by energy-analysing the scattered beam. The use of different single crystal types on the analyser array alters the instrument resolution and thus the observable time window. The most used crystal types are silicon ([111], [311] reflections) and pyrolytic graphite ([002], [004] reflections). (iii) Illustrative 'bird's eye view' of an indirect geometry time of flight instrument, which uses a chopper system to filter the neutron flux distribution generated by the source and define ΔE_i.

Doppler shift. The incident beam is therefore no longer wholly monochromatic but exhibits a finite wavelength, or energy, spread, ΔE_i. It should be mentioned that instruments that function in this manner do not operate in the classic ToF mode described earlier. Instead, detected neutrons are assigned to energy channels by the instrument electronics; the channel used depending upon the state (velocity characteristics) of the Doppler drive the moment the detected neutron was Bragg reflected towards the sample. It should also be mentioned that Doppler shifting is not the only mechanism used to generate ΔE_i. The IN13 instrument at the Institut Laue-Langevin, for example, uses thermal expansion (by heating and cooling its monochromator) to scan the crystal's lattice parameter and thus modify the Bragg condition. In contrast, indirect geometry instruments that employ the ToF method use band width choppers (neutron absorbing discs with wide neutron-transparent apertures) to transmit a narrow band of incident energies to the sample position. It should be mentioned that the reactor based instrument, IN16B, while historically a Doppler-based instrument, can be configured to operate in ToF mode.[11]

In terms of defining E_f, both pulsed and reactor source indirect geometry instruments do so by energy-analysing the scattered neutron beam. Energy analysis is achieved by placing a large (to improved detected flux) array of single crystals (typically silicon or pyrolytic graphite) in the path of the scattered neutron beam, between sample and detector. Only those elastically scattered, or energy exchanged, neutrons that leave the sample with a wavelength that satisfies the Bragg condition are reflected towards the detector.

As with the direct geometry instrument, it is worth considering the sources that limit the sensitivity of an indirect geometry spectrometer. The energy resolution of a backscattering instrument viewing a pulsed neutron source can be written in terms of,[12]

$$\left(\frac{\Delta E}{2E}\right)^2 = \left(\frac{\Delta t}{t}\right)^2 + \left(\frac{\Delta L}{L}\right)^2 + \left(\frac{\Delta d}{d}\right)^2 + \left(\cot(\theta)\Delta(\theta)\right)^2 \qquad (3.3)$$

i.e. a total ToF uncertainty (Δt), a total instrument length uncertainty $(\Delta L, L = L_1 + L_2)$, an uncertainty in the lattice spacing of the analyser crystal used to energy analyse the scattered beam (Δd) and beam divergence $(\Delta(\theta))$. The terms can be added in quadrature if they are assumed to be statistically independent and can be further grouped into primary and secondary spectrometer contributions associated with both E_i and E_f.

For a steady state research reactor based instrument, however, then eqn (3.3) reduces to,[13]

$$\left(\frac{\Delta E_f}{2E_f}\right) = \left(\frac{\Delta d}{d}\right) + \left(\cot((\theta))\Delta(\theta)\right) \qquad (3.4)$$

One can see that to achieve the highest energy resolutions, indirect geometry instruments (i) must operate such that the energy-analysed beam is Bragg scattered through 180° and (ii) require negligible timing uncertainty. To achieve these conditions, therefore, the detector array needs to be located directly behind the sample position with the instrument sited at a research reactor. Spectrometers configured in this manner are known as backscattering instruments. It should be noted that such backscattering spectrometers do afford high energy resolution, and as such allow access the nanosecond time regime.

However, they do so at the expense of a sizeable energy transfer window. In contrast, and while perhaps not as sensitive, indirect geometry ToF spectrometers viewing pulsed sources allow access to wide dynamic ranges and can offer sizeable improvements in neutron flux. Such instruments also operate with θ close to 180° such that their detector array is set slightly below the sample position. As a result, they are referred to as near-backscattering spectrometers.

3.2.4 Neutron Spin Echo (NSE)

The technique of neutron spin echo considerably extends the temporal and spatial landscape available to quasi-elastic spectroscopy. Not only does it allow access to motions occurring on the tens to hundreds of nanoseconds, but, as Figure 3.1 shows, covers a spatial range similar to that probed by the technique of small angle neutron scattering (SANS). As a result, NSE can be used as a complementary tool to probe the dynamics associated with structural features observed in $S_c(\mathbf{Q})$.

The technique was pioneered by F Mezei[14] and measures variations in the velocity of the neutron during the scattering process by tracking changes in the spin polarisation of the neutron beam. A schematic diagram showing a typical spin echo set up is shown in Figure 3.6. In truth, the underlying operating principles of NSE instrumentation require time and experience to master. You would not, therefore, be expected to set up or configure an NSE instrument without the help of neutron facility staff. However, it is a hope that the following description of the spectrometer will aid understanding of this measurement method.

The technique can be described as follows. A velocity selector (1) and super-mirror neutron polariser (2) assembly are used to longitudinally spin-polarise a broadly monochromatic incident beam of neutrons with $\Delta\lambda_i/\lambda_i$ values between ~10 and −20% (*e.g.* the NSE instrument at NCNR often uses ~17%). The polarised beam is rotated by $\pi/2$ radians using a $\pi/2$-flipper (3) into the *x*-direction before travelling along the (*z*-) axis of a solenoid (S_1) of length $= L_1$ and longitudinal field strength, B_1 (4). The neutrons enter S_1 with their spin polarisation perpendicular to the field direction and consequently Larmor precess in the *x*–*y* plane. The polarisation component of a neutron of velocity, v_1, along the *x* direction perpendicular to B_1 is given by,

$$P_x = <\cos(\varphi_1)> = \int f(v)\cos\left(\frac{\gamma_L \int B_1\, dl}{v}\right) dv \qquad (3.5)$$

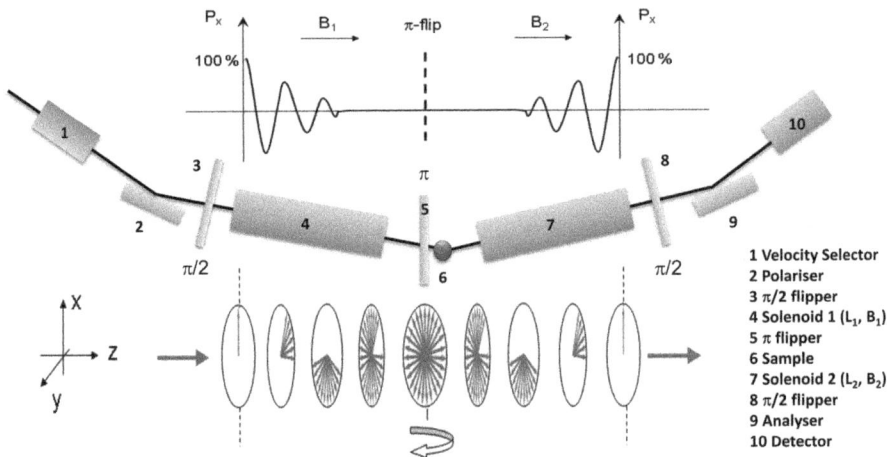

Figure 3.6 A schematic of the neutron spin echo method and spectrometer components.

Here, φ_1 is the precession angle (in radians) accumulated over the distance L_1, γ_L is the gyro-magnetic ratio of the neutron and $f(v)$ is the velocity distribution function. Since the incident neutron beam is broadly monochromatic, the precessing spins, associated with neutrons of differing velocities, will de-phase and $P_x = \langle\cos(\varphi_1)\rangle$ will tend to zero at the exit of the first solenoid (aka arm).

Before passing through the second solenoid (S_2, length $= L_2$, longitudinal field strength $= B_2$) (7) the neutron spins are flipped again through π radians using a π-flipper (5). The de-phased neutron spins again precess in the x–y plane of S_2 but, because of the inversion at the π-flipper, they are now unwinding the phase spread accumulated in S_1. If φ_2 is the total precession angle after distance L_2 then the total precession angle, φ_T, over a total flight path $L_T = L_1 + L_2$ is,

$$\varphi_T\left(v_1,v_2\right)=\varphi_2-\varphi_1=\gamma_L\left(\frac{\int B_2\,\mathrm{d}l}{v_2}-\frac{\int B_1\,\mathrm{d}l}{v_1}\right) \tag{3.6}$$

If the two solenoids are of equal lengths, and the guide field gradients are equivalent, then, for an elastically scattering material, full beam polarisation will be recovered at the exit of the second arm (S_2). This is known as the 'echo' condition. However, for a quasi-elastic scattering event, the velocity of a scattered neutron will change and $v_2 = v_1 \pm \delta v$. Quasi-elastic scattering, therefore, will result in an unequal number of neutron precessions in the two solenoids and thus loss of spin polarisation.

The energy transfer associated with the quasi-elastic process is related to the velocity of the incident and scattered neutrons *via* $\Delta E = \Delta \hbar \omega = E_1 - E_2 = \tfrac{1}{2} m_n (v_2^2 - v_1^2) = \hbar \omega(v_1, v_2)$. Therefore the total Larmor precession angle, $\varphi_T(v_1, v_2)$, and hence spin polarisation, can be related to the energy transfer, $\hbar \omega(v_1, v_2)$. In the quasi-elastic limit, $v_2 = v_1 \pm \delta v$, $\Delta E = \hbar \omega = m v \delta v$ and therefore $v_2 = v_1 \pm \hbar \omega / m v$. Incorporating this into eqn (3.6), and expanding to the first order in ω, the total accumulated precession angle, φ_T, becomes,

$$\varphi_T = \frac{\gamma_L (B_1 L_1 - B_2 L_2)}{v_1} + \frac{\gamma_L B_2 L_2 \hbar \omega}{m v_1^3} \tag{3.7}$$

At the echo condition, $B_1 L_1 = B_2 L_2$ and the accumulated precession angle is,

$$\varphi_T' = \frac{\gamma_L B_2 L_2 \hbar \omega}{m v_1^3} = t_F \omega \tag{3.8}$$

where t_F is a constant of proportionality that has the units of time. t_F is directly proportional to the line integrals $B_1 L_1$ and $B_2 L_2$ and is also directly proportional to λ^3, where λ is the incident wavelength. Therefore, for quasi-elastic scattering, the spin polarisation close to the elastic echo condition will decrease to a value of $\cos(\varphi_T')$, or $\cos(\omega t_F)$. The scattering function is a measure of the distribution of precession angles in the scattered beam since it describes the probability that a neutron is scattered with an energy change, $\hbar \omega$. The net polarisation of the beam can be averaged over all quasi-elastic scattering processes to give,

$$P_z(\mathbf{Q}, t_F) = \langle \cos(\omega t_F) \rangle = \int S_{\text{NSE}}(\mathbf{Q}, \omega) \cos(\omega t_F) \, d\omega = I_{\text{NSE}}(\mathbf{Q}, t_F) \tag{3.9}$$

which represents the Fourier transform of $S_{\text{NSE}}(\mathbf{Q}, \omega)$ with respect to ω and where, in the quasi-elastic limit, the Fourier time, t_F, is equivalent to real time.

It should be noted that spin incoherent scattering flips the neutron spins during scatter with a 66% probability, thus,

$$S_{\text{NSE}}(\mathbf{Q}, \omega) = \left[S_c(\mathbf{Q}, \omega) - \frac{1}{3} S_i(\mathbf{Q}, \omega) \right] \tag{3.10}$$

Hence, the dynamic structure factor probed by NSE is different from that probed on other spectrometers, which is usually given by the sum of the coherent and incoherent contributions. The presence of incoherent scattering introduces a depolarisation of the beam that is not related to the dynamics of the system. For this reason, in conjunction with the echo measurement, it is customary to perform a polarised diffraction measurement to determine the integral of $S_{NSE}(\mathbf{Q},\omega)$ over all energy, *i.e.* namely the static structure factor,

$$I_{NSE}(\mathbf{Q},0) = I_{NSE}(\mathbf{Q}) = S_{NSE}(\mathbf{Q}) = \left[S_c(\mathbf{Q}) - \frac{1}{3} S_i(\mathbf{Q}) \right] \qquad (3.11)$$

This is carried out by switching off the $\pi/2$ flippers and measuring with the π-flipper first on, and then off. Intuitively, one can appreciate that with the $\pi/2$ flippers off there is no precession and therefore zero Fourier time. The final result for a given t_F is simply a measurement of the normalised intermediate scattering law,

$$I_{NSE}(\mathbf{Q}, t_F)/I_{NSE}(\mathbf{Q}) \qquad (3.12)$$

It should also be noted that materials that exhibit an incoherent scattering contribution can prove disadvantageous for the NSE method. The measured normalised intermediate scattering law may be a complicated combination that cannot be decomposed if only NSE is employed. However, in pure cases, *i.e.* when there is an overwhelming intensity contribution due to protons, NSE can be used to measure an incoherent spectrum,

$$\frac{I_{NSE}(\mathbf{Q},t)}{I_{NSE}(\mathbf{Q})} = \frac{I_c(\mathbf{Q},t) - \frac{1}{3} I_s(\mathbf{Q},t)}{I_c(\mathbf{Q},0) - \frac{1}{3} I_s(\mathbf{Q},0)} \approx \frac{I_s(\mathbf{Q},t)}{I_s(\mathbf{Q},0)} \qquad (3.13)$$

In principle then, NSE can access both coherent and incoherent signals. However, the incoherent contribution is usually weak and, as such, NSE is more suited to the study of collective motion(s). In practice, the polarised diffraction measurement provides the relative intensity of the coherent and incoherent scattering; however, the two dynamic contributions cannot be easily separated experimentally using a standard NSE spectrometer set-up.

3.3 Polarisation Analysis Options on Direct and Indirect Instruments

The analysis of self or collective motions should ideally be performed by modelling $S(\mathbf{Q},\omega)$ from a system that exhibits purely incoherent or coherent scattering, *i.e.* $S(\mathbf{Q},\omega) = S_i(\mathbf{Q},\omega)$ or $S_c(\mathbf{Q},\omega)$. As discussed, for strongly hydrogenous materials we approximate that the measured signal is dominated by the sizeable incoherent scattering cross section of the hydrogen atom, and thus predominately arises from $S_i(\mathbf{Q}, \omega)$. However, for systems that are not protonated, or for materials where deuteration has been used to 'mask' particular components, $S_i(\mathbf{Q},\omega)$ and $S_c(\mathbf{Q},\omega)$ may coexist and, subsequently, complicate data interpretation. Of course, should coherence manifest itself as Bragg lines at well-defined Q values in a quasi-elastic scattering data set (*i.e.* a signature of well-defined long range crystallographic order) then the simplest approach is to exclude the detectors in which the Bragg contamination occurs from the analysis; this of course being done at the expense of Q, or length scale information. Unfortunately, liquid, amorphous or fine powder samples exhibit predominantly diffuse static structure factors, $S(\mathbf{Q})$ (*i.e.* no well-defined Bragg lines) that can appear like a background component and span an instrument's entire momentum transfer range.

When not using NSE instrumentation, coherent scattering contributions are particularly difficult to isolate. In practice, installation of additional apparatus on a beamline (*e.g.* super-mirror polarisers, ^{3}He spin filters) is required to analyse the spin polarity of the scattered beam. An overview of the measurements necessary to separate the two components can be found in Gaspar *et al.*[15] Direct and indirect geometry instrumentation that offers such polarisation analysis (PA) capability is not routinely available. In fact, and perhaps counter intuitively, the first instrument to show possibility in this area was the instrument D7 at the Institut Laue-Langevin.[16] While a diffuse scattering diffractometer by design, the instrument can be converted to a low resolution QENS spectrometer. However, developments on the direct geometry instrument, LET[17] at ISIS, UK, as well as MACS[18] (Multi Axis Crystal Spectrometer) at NIST, USA, using ^{3}He neutron spin filter technologies are helping to extend opportunities in this area at both research reactor and pulsed facilities.

3.4 International QENS Instrument Landscape

For reference, Table 3.2 summarises those neutron instruments most associated with QENS studies housed at currently operating neutron science facilities. The Q ranges and energy resolutions quoted are those associated with standard operating configurations as determined from the given references and on-line instrument resources.

Table 3.2 A summary of those neutron instruments most associated with QENS studies at currently operating neutron science facilities. The Q ranges and energy resolutions quoted are those associated with standard operating configurations as determined from the given references and online instrument resources. It should be noted that instruments perhaps more aligned with INS studies may still be suitable for QENS investigations.[19]

Host facility	Instrument name	Instrument class	Q range ($\mathring{A}^{-1}$) @ $\Delta\hbar\omega = 0$	Elastic line energy resolution Γ^{res} (μeV, FWHM), $\Delta E/E_i$ (%) or t_F (ns)
		Spallation/pulsed sources		
ISIS[a] Harwell, UK. Commenced operation: 1985 Power: 0.16 MW	**IRIS**[20]	ToF, indirect	$0.42 < Q < 1.85$ $0.84 < Q < 3.70$	17.5 μeV, PG(002) 54.5 μeV, PG(004)
	OSIRIS[21]	ToF, indirect	$0.18 < Q < 1.8$ $0.37 < Q < 3.6$	25.4 μeV, PG(002) 99 μeV, PG(004)
	LET[8]	ToF, direct	$Q_{min} = 0.032 \times \sqrt{E_i}$ $Q_{max} = 1.32 \times \sqrt{E_i}$	20 μeV @ $E_i = 1$ meV 500 μeV @ $E_i = 20$ meV
SINQ, PSI[b] Villigen PSI, Switzerland. Commenced operation: 1996 Power: 1 MW	**FOCUS**[22]	ToF, direct	$0.55 < Q < 5.7$ ($E_i = 20.45$ meV) $0.18 < Q < 1.89$ ($E_i = 2.27$ meV)	45 μeV @ $E_i = 2.27$ meV 1.7 meV @ $E_i = 20.45$ meV
Note: SINQ is a quasi-continuous source				
SNS, ORNL[c] Oak Ridge, USA. Commenced operation: 2006 Power: 1.4 MW	**BASIS**[13]	ToF, indirect	$0.2 < Q < 2.0$ $0.4 < Q < 3.8$	3.5 μeV, Si(111) 15 μeV, Si(311)
	CNCS[23]	ToF, direct	$0.05 < Q < 10$	15 μeV @ 1 meV 1000 μeV @ 20 meV
	SNS-NSE[24]	ToF, NSE	$0.03 < Q < 2.0$	$0.001 < t_F < 300$ ns For 300 ns, $\lambda_i = 14$–17 $\mathring{A}$

J-PARC, KEK/JAEA[d,19] Tokai, Japan. Commenced operation: 2008 Power: 1 MW			
DNA[5]	ToF, indirect	$0.08 < Q < 1.86$	Si(111) 1.4 µeV, slits 1 cm, sample 1 cm high; 2.2 µeV, slits 1 cm, sample 3 cm high; 3.1 µeV, slits 3 cm, sample 3 cm high
		$2.46 < Q < 3.4$	Si(311) 12 µeV, slits 1 cm, sample 3 cm height
Note: Meize and NRSE make up VIN ROSE – The Village of Neutron ResOnance Spin Echo Spectrometers			
AMATERAS[6,25,26]	ToF, direct	$0.06 < Q < 1.55 \ (E_i = 1.7 \text{ meV})$ $0.3 < Q < 7.7 \ (E_i = 42 \text{ meV})$	$\Delta E/E_i > 1\%$ @ $E_i = 20$ meV, $E_i = 1$ to 80 meV
MIEZE[19,27]	ToF, NSE	$0.2 < Q < 3.5$	$0.001 < t_F < 2$ ns
NRSE[25]	ToF, NSE	$0.02 < Q < 0.65$	$0.1 < t_F < 100$ ns
MIRACLES[28]	ToF, indirect	$0.2 < Q < 1.8$	2.5–30 µeV, Si(111)
ESS, ERIC[e] Lund, Sweden. Construction started: 2014 Power: 5 MW			
CSPEC	ToF, direct	$0.27 < Q < 5.9 \ (2 \text{ Å})$ $0.11 < Q < 2.3 \ (5 \text{ Å})$	$\Delta E/E_i \sim 1\text{–}5\%$. 1.5% @ 4.0 Å

Medium-flux reactors ($10^{14} < flux < 10^{15}$ neutrons per second per cm^2)

NCNR, NIST[f] Gaithersburg, USA. Commenced operation: 1967 Power: 20 MW			
HFBS[10]	Doppler, indirect	$0.25 < Q < 1.75$	0.85 µeV, Si(111)
DCS[29]	ToF, direct	$0.3 < Q < 6.5 \ (1.8 \text{ Å})$	950 µeV @ 1.8 Å
		$0.1 < Q < 1.2 \ (10 \text{ Å})$	17 µeV @ 10 Å
CHRNS NSE[30]	NSE	$0.02 < Q < 0.6 \ (15 \text{ Å})$ $0.06 < Q < 1.8 \ (5 \text{ Å})$	$0.003 < t_F < 200$ ns

(continued)

Table 3.2 (continued)

Host facility	Instrument name	Instrument class	Q range (Å^{-1}) @ $\Delta\hbar\omega = 0$	Elastic line energy resolution Γ_{res} (μeV, FWHM), $\Delta E/E_i$ (%) or t_F (ns)
FRM II[g] Garching, Germany. Commenced operation: 2004 Power: 20 MW	**SPHERES**[31]	Doppler, indirect	$0.2 < Q < 1.8$	0.65 μeV, Si(111)
	TOFTOF[32]	ToF, direct	$0.8/\lambda_i < Q < 11.8/\lambda_i$	3000 μeV @ 1.6 Å 100 μeV @ 5 Å
	RESEDA[33,34]	NRSE, MIEZE	$Q_{max} = 2.0 \ (\lambda = 4.5$ Å$)$	$0.0001 < t_F < 20$ ns @ $\lambda_i = 8$ Å
	J-NSE[35,36] 'Phoenix'	NSE	$0.02 < Q < 1.7$	$0.002 \ (4.5$ Å$) < t_F < 500$ ns (16 Å)
ANSTO[h] Sydney, Australia. Commenced operation: 2007 Power: 20 MW	**EMU**[37]	Doppler, indirect	$0.1 < Q < 1.95$	0.95 μeV, Si(111)
	PELICAN[38]	ToF, direct	$0.08 < Q < 4.5$	50–350 μeV (~2.5% of E_i) 14.2 meV < E_i < 2.1 meV

High flux reactors (flux > 10^{15} neutrons per second per cm^2)

Host facility	Instrument name	Instrument class	Q range (Å^{-1}) @ $\Delta\hbar\omega = 0$	Elastic line energy resolution Γ_{res} (μeV, FWHM), $\Delta E/E_i$ (%) or t_F (ns)
ILL[i] Grenoble, France. Commenced operation: 1972 Power: 58.3 MW	**IN13**[39]	Thermal, indirect	$0.2 < Q < 4.9$	8.0 μeV, CaF$_2$(422)
	IN16B[40]	Doppler, indirect	$0.1 < Q < 1.8$ $0.7 < Q < 3.5$	0.75 μeV Si(111) 2.0 μeV Si(311)
		ToF, indirect (BATS option)[11]	$0.2 < Q < 1.8$	1.5 μeV (high resolution), Si(111) 8.6 μeV (high flux), Si(111)
	IN5[41]	ToF, direct	$0.5/\lambda_i < Q < 11.6/\lambda_i$	1.2 meV @ 2 Å 100 μeV @ 5 Å 11 μeV @ 9 Å

IN6 – SHARP[42]	ToF, direct	$0.27 < Q < 2.6$ ($E_i = 4.8$ meV) $0.18 < Q < 1.8$ ($E_i = 2.33$ meV)	170 μeV @ 4.8 meV 120 μeV @ 3.83 meV 70 μeV @ 3.11 meV 50 μeV @ 2.33 meV
IN15[43]	NSE	$0.01 < Q < 0.42$ (17 Å)	$0.03 < t_F < 1000$ ns (17 Å)
WASP[44]	NSE	$0.03 < Q < 3.0$ (4 Å) $0.02 < Q < 1.2$ (10 Å)	$0.002 < t_F < 3$ ns (4 Å) $0.03 < t_F < 50$ ns (10 Å)

[a]**ISIS**[45] The ISIS Pulsed Neutron and Muon Source. Part of the UK's Science and Technology Facilities Council.
[b]**SINQ, PSI**[46] The Swiss Spallation Neutron Source located at the Paul Scherrer Institut, Switzerland.
[c]**SNS, ORNL**[47] The Spallation Neutron Source located at the Oak Ridge National Laboratory, Tennessee, USA.
[d]**J-PARC**[48] , **KEK/JAEA**[49] The Japan Proton Accelerator Research Complex, which is a joint project between The High Energy Accelerator Research Organization (KEK) and the Japan Atomic Energy Agency (JAEA).
[e]**ESS, ERIC**[50] The European Spallation Source, which is a European Research Infrastructure Consortium.
[f]**NCNR, NIST**[51] The NIST Center for Neutron Research, part of the National Institute of Standards and Technology (NIST).
[g]**FRM II**[52] Forschungs-Neutronenquelle Heinz Maier-Leibnitz (Research Neutron Source Heinz Maier-Leibnitz) operated by Technische Universität München, Germany.
[h]**ACNS, ANSTO**[53] The Australian Centre for Neutron Scattering, part of the Australian Nuclear Science and Technology Organisation (ANSTO), which hosts the 20 megawatt (MW) Open-Pool Australian Light-water reactor (OPAL).
[i]**ILL**[54] Institut Laue-Langevin, Grenoble, France, governed by three founding Associate countries and ten Scientific Member countries.

Finally, the following instruments are no longer operational but are referenced due to their contribution to the field:

- **NEAT**[55] A time-of-flight spectrometer at the BER-II reactor, Helmholtz-Zentrum, Berlin, Germany.
- **QENS**[56] A medium resolution, inverted geometry, time-of-flight quasi-elastic and inelastic spectrometer at the Intense Pulsed Neutron Source (IPNS), Argonne National Lab (ANL), USA.
- **Mibémol**[57] A disk-chopper time-of-flight instrument at the ORPHEE reactor, Laboratoire Léon Brillouin (LLB); a French laboratory supported jointly by the Commissariat à l'Energie Atomique (CEA) and the Centre National de la Recherche Scientifique (CNRS).
- **IN11A, IN11C**[58] A: 2D multi-detector setup. C: A wide angle detector setup. IN11 was the first neutron spin-echo spectrometer operated at the Institut Laue-Langevin, Grenoble, France.

References

1. J. S. Gardner, G. Ehlers, A. Faraone and V. García Sakai, *Nat. Rev. Phy.*, 2020, 103–116.
2. R. J. Newport, B. D. Rainford and R. Cywinski, *Neutron Scattering at a Pulsed Source*, A. Hilger, Bristol, England; Philadelphia, PA, USA, 1988.
3. https://www.mlz-garching.de/englisch/neutron-research/neutron-source.html.
4. E. Mamontov, M. Zamponi, S. Hammons, W. S. Keener, M. Hagen and K. W. Herwig, *Neutron News*, 2008, **19**, 22–24.
5. N. Takahashi, K. Shibata, Y. Kawakita, K. Nakajima, Y. Inamura, T. Nakatani, H. Nakagawa, S. Fujiwara, T. J. Sato, I. Tsukushi, F. Mezei, D. A. Neumann, H. Mutka and M. Arai, *J. Phys. Soc. Jpn.*, 2011, **80**, SB007.
6. K. Nakajima, S. Ohira-Kawamura, T. Ktxucm, M. Nakamura, R. Kajimoto, Y. Inamura, N. Takahashi, K. Atzawa, K. Suzuya, K. Shibata, T. Nakatani, K. Soyama, R. Maruyama, H. Tanaka, W. Kambara, T. Iwahasmi, Y. Itoh, T. Osakabe, S. Wakimoto, K. Kakurai, F. Maekawa, M. Harada, K. Oikawa, R. E. Lechner, F. Mezei and M. Arai, *J. Phys. Soc. Jpn.*, 2011, **80**, SB028.
7. C. G. Windsor, *Pulsed Neutron Scattering*, Taylor and Francis, London, 1981.
8. R. I. Bewley, J. W. Taylor and S. M. Bennington, *Nucl. Instrum. Methods Phys. Res., Sect. A*, 2011, **637**, 128–134.
9. B. Frick and M. Gonzalez, *Physica B*, 2001, **301**, 8–19.
10. A. Meyer, R. M. Dimeo, P. M. Gehring and D. A. Neumann, *Rev. Sci. Instrum.*, 2003, **74**, 2759.
11. M. Appel, B. Frick and A. Magerl, *Sci. Rep.*, 2018, **8**, 13580.
12. F. Demmel and K. H. Andersen, *Meas. Sci. Technol.*, 2008, **19**, 034021.
13. E. Mamontov and K. W. Herwig, *Rev. Sci. Instrum.*, 2011, **82**, 085109.
14. F. Mezei, *Z. Phys.*, 1972, **255**, 146–150.

15. A. M. Gaspar, S. Busch, M. S. Appavou, W. Haeussler, R. Georgii, Y. X. Su and W. Doster, *Biochim. Biophys. Acta, Proteins Proteomics*, 2010, **1804**, 76–82.
16. J. R. Stewart, P. P. Deen, K. H. Andersen, H. Schober, J. F. Barthelemy, J. M. Hillier, A. P. Murani, T. Hayes and B. Lindenau, *J. Appl. Crystallogr.*, 2009, **42**, 69–84.
17. G. J. Nilsen, J. Kosata, M. Devonport, P. Galsworthy, R. I. Bewley, D. J. Voneshen, R. Dalgliesh and J. R. Stewart, in *International Conference on Polarised Neutrons for Condensed Matter Investigations*, 2017, vol. 862.
18. W. Chen, S. Watson, Y. Qiu, J. A. Rodriguez-Rivera and A. Faraone, *Phys. B*, 2019, **564**, 166–171.
19. H. Seto, S. Itoh, T. Yokoo, H. Endo, K. Nakajima, K. Shibata, R. Kajimoto, S. Ohira-Kawamura, M. Nakamura, Y. Kawakita, H. Nakagawa and T. Yamada, *Biochim. Biophys. Acta*, 2017, **1861**, 3651–3660.
20. C. J. Carlile and M. A. Adams, *Physica B*, 1992, **182**, 431–440.
21. M. T. F. Telling, S. I. Campbell, D. Engberg, D. Martín y Marero and K. H. Andersen, *Phys. Chem. Chem. Phys.*, 2005, **7**, 1255–1261.
22. S. Janssen, J. Mesot, L. Holitzner, A. Furrer and R. Hempelmann, *Physica B*, 1997, **234**, 1174–1176.
23. G. Ehlers, A. A. Podlesnyak, J. L. Niedziela, E. B. Iverson and P. E. Sokol, *Rev. Sci. Instrum.*, 2011, **82**, 085108.
24. M. Ohl, M. Monkenbusch, N. Arend, T. Kozielewski, G. Vehres, C. Tiemann, M. Butzek, H. Soltner, U. Giesen, R. Achten, H. Stelzer, B. Lindenau, A. Budwig, H. Kleines, M. Drochner, P. Kaemmerling, M. Wagener, R. Moller, E. B. Iverson, M. Sharp and D. Richter, *Nucl. Instrum. Methods Phys. Res., Sect. A*, 2012, **696**, 85–99.
25. K. Nakajima, Y. Kawakita, S. Itoh, J. Abe, K. Aizawa, H. Aoki, H. Endo, M. Fujita, K. Funakoshi, W. Gong, M. Harada, S. Harjo, T. Hattori, M. Hino, T. Honda, A. Hoshikawa, K. Ikeda, T. Ino, T. Ishigaki, Y. Ishikawa, H. Iwase, T. Kai, R. Kajimoto, T. Kamiyama, N. Kaneko, D. Kawana, S. Ohira-Kawamura, T. Kawasaki, A. Kimura, R. Kiyanagi, K. Kojima, K. Kusaka, S. Lee, S. Machida, T. Masuda, K. Mishima, K. Mitamura, M. Nakamura, S. Nakamura, A. Nakao, T. Oda, T. Ohhara, K. Ohishi, H. Ohshita, K. Oikawa, T. Otomo, A. Sano-Furukawa, K. Shibata, T. Shinohara, K. Soyama, J.-i. Suzuki, K. Suzuya, A. Takahara, S.-i. Takata, M. Takeda, Y. Toh, S. Torii, N. Torikai, L. N. Yamada, T. Yamada, D. Yamazaki, T. Yokoo, M. Yonemura and H. Yoshizawa, *Quantum Beam Sci.*, 2017, **1**, 9–64.
26. R. Kajimoto, T. Yokoo, M. Nakamura, Y. Kawakita, M. Matsuura, H. Endo, H. Seto, S. Itoh, K. Nakajima and S. Ohira-Kawamura, *Phys. B*, 2019, **562**, 148–154.
27. Y. Kawabata, M. Hino, M. Kitaguchi, H. Hayashida, S. Tasaki, T. Ebisawa, D. Yamazaki, R. Maruyama, H. Seto, M. Nagao and T. Kanaya, *Phys. B*, 2006, **385–86**, 1122–1124.
28. N. Tsapatsaris, R. E. Lechner, M. Marko and H. N. Bordallo, *Rev. Sci. Instrum.*, 2016, **87**, 085118.
29. J. R. D. Copley and J. C. Cook, *Chem. Phys.*, 2003, **292**, 477–485.
30. N. Rosov, S. Rathgeber and M. Monkenbusch, in *Scattering from Polymers: Characterization by X-Rays, Neutrons, and Light*, 2000, vol. 739, pp. 103–116.
31. J. Wuttke, A. Budwig, M. Drochner, H. Kammerling, F. J. Kayser, H. Kleines, V. Ossovyi, L. C. Pardo, M. Prager, D. Richter, G. J. Schneider, H. Schneider and S. Staringer, *Rev. Sci. Instrum.*, 2012, **83**, 075109.
32. T. Unruh, E. Neuhaus and W. Petry, *Nucl. Instrum. Methods Phys. Res., Sect. A*, 2008, **585**, 201.
33. C. Franz, O. Soltwedel, C. Fuchs, S. Säubert, F. Haslbeck, A. Wendl, J. K. Jochum, P. Böni and C. Pfleiderer, *Nucl. Instrum. Methods Phys. Res., Sect. A*, 2019, **939**, 22–29.
34. W. Haussler, B. Gohla-Neudecker, R. Schwikowski, D. Streibl and P. Boni, *Phys. B*, 2007, **397**, 112–114.

35. S. Pasini, O. Holderer, T. Kozielewski, D. Richter and M. Monkenbusch, *Rev. Sci. Instrum.*, 2019, **90**, 043107.
36. O. Holderer, M. Monkenbusch, R. Schatzler, H. Kleines, W. Westerhausen and D. Richter, *Meas. Sci. Technol.*, 2008, **19**, 034022.
37. N. R. de Souza, A. Klapproth and G. N. Iles, *Neutron News*, 2016, **27**, 20–21.
38. D. H. Yu, R. Mole, T. Noakes, S. Kennedy and R. Robinson, *J. Phys. Soc. Jpn.*, 2013, **82**, SA027.
39. F. Natali, M. Bee, A. Deriu, C. Mondelli, L. Bove, C. Castellano and S. Labbe-Lavigne, *Phys. B*, 2004, **350**, E819–E822.
40. B. Frick, H. N. Bordallo, T. Seydel, J. F. Barthelemy, M. Thomas, D. Bazzoli and H. Schober, *Phys. B*, 2006, **385–86**, 1101–1103.
41. J. Ollivier, H. Mutka and L. Didier, *Neutron News*, 2010, **21**, 22–25.
42. J. Ollivier and J.-M. Zanotti, *Diffusion Inélastique des Neutrons pour l'Étude des Excitations dans la Matière Condensée Albé, France, May 2008*, 2010, vol. 23–27.
43. P. Schleger, B. Alefeld, J. F. Barthelemy, G. Ehlers, B. Farago, P. Giraud, C. Hayes, A. Kollmar, C. Lartigue, F. Mezei and D. Richter, *Physica B*, 1997, **241**, 164–165.
44. P. Fouquet, G. Ehlers, B. Farago, C. Pappas and F. Mezei, *J. Neutron Res.*, 2007, **15**, 39–47.
45. http://www.isis.stfc.ac.uk/.
46. https://www.psi.ch/sinq/.
47. https://neutrons.ornl.gov/instruments.
48. http://j-parc.jp/researcher/MatLife/en/instrumentation/ns.html.
49. https://www.kek.jp/en/Facility/ACCL/J-PARC/.
50. https://europeanspallationsource.se/.
51. https://www.nist.gov/ncnr.
52. https://www.mlz-garching.de/englisch/neutron-research/neutron-source.html.
53. https://www.ansto.gov.au/research/facilities/australian-centre-for-neutron-scattering.
54. http://www.ill.eu/.
55. R. E. Lechner, *Neutron News*, 1996, **7**, 9–11.
56. K. F. Bradley, S. H. Chen, T. O. Brun, R. Kleb, W. A. Loomis and J. M. Newsam, *Nucl. Instrum. Methods Phys. Res., Sect. A*, 1988, **270**, 78–89.
57. J.-M. Zanotti, S. Combet, S. Klimko, S. p. Longeville and F. d. r. Coneggo, *Neutron News*, 2011, **22**, 24–27.
58. B. Farago, *Physica B*, 1997, **241**, 113–116.

Part 2

Measurement

4 Facility Access

In this chapter we will consider:

- How to gain access to neutron facilities and instruments.
- Ingredients for writing successful facility access applications.
- The application review process.

4.1 You Have an Idea...

You have a well-defined research problem, a suitably characterised sample and now an 'interesting' hypothesis to test based on experimental evidence that suggests a need for neutron spectroscopy. So, how do you gain access to neutron instrumentation?

This chapter outlines basic facility access routes and considers the key ingredients for writing successful access applications. However, it should be stressed that for the most up-to-date information, and to understand the subtle differences between facilities regarding access protocols, it is always best to speak to facility staff and/or refer to a specific facility's online resource.

Neutron facilities operate online access systems whereby researchers are invited to upload expressions of interest (aka beam time proposals). There are usually two calls for proposals per year (usually spring and autumn) but this can depend upon, for example, facility upgrade schedules. Broadly speaking, neutron instruments are free at the point of access for academic and industry researchers provided the results from experiments are published in the public domain.

A Practical Guide to Quasi-elastic Neutron Scattering
By Mark T. F. Telling
© Mark T. F. Telling 2020
Published by the Royal Society of Chemistry, www.rsc.org

However, facilities may offer fully confidential, 'fast-tracked' use of the instruments for industrial and commercial customers.

Most facilities adopt online templates and submission tools to capture general information about an experiment, *i.e.* the experimental conditions required, sample composition and safety issues, grant information associated with the research, associated PhD studentships, *etc.* However, and more importantly, these templates are accompanied by a written 2–3 side scientific summary of the work proposed.

All submitted proposals are peer reviewed. However, with little additional information to guide a review panels' decision, success rides on the clarity of your scientific summary. Furthermore, with access to neutron facilities in demand, acceptance can also depend upon other factors, some beyond your control: facility operational periods, instrument oversubscription, high-risk experiments *vs.* guaranteed publications, review panel fatigue and national balance.

4.2 Before Writing Your Proposal

While time consuming, effort spent pre-planning will help you contextualise your research. It is therefore advisable to:

- Review the scientific literature. This will help you state clearly how your experiment fits into the broader scheme and how it will advance the field.
- Check whether similar experiments have been performed. If so, what lessons can be learned?
- Collate evidence that demonstrates that neutron scattering is the right technique to answer your hypothesis(es). As will be discussed (Section 4.4), and if available at the facility of your choice, presenting results from an express (aka Easy Access or Xpress) study is most beneficial here.
- Identify who can be listed as the main contact (aka Principle Investigator) on any access application. You may find (if you are a student) that your supervisor will need to upload the documentation and oversee all subsequent communication. For researchers not based in a facility's associate/partner-country, be sure that you can apply for access without a partner-country collaborator.

- Monitor resources for neutron-related news, activities and developments:
  - colleagues
  - facility web pages
  - scientific literature
  - facility annual reports
  - www.neutronsources.org
  - facility staff (especially the neutron instrument responsible)
  - neutron community mailing lists (visit www.neutronsources.org).

4.3   Which Facility Should You Choose?

For those not influenced by personal connection, supervisor preference and/or existing collaborations, it is worth identifying those facilities that offer instruments that align with your research needs. The choice of facility will then be honed once you have then noted which of these spectrometers afford suitable performance specifications and sample environment apparatus; this is especially important if your research requires your sample to be subject to the experimental extremes listed in Table 1.2. However, in truth, geographical location and funding support will greatly influence your decision. Other factors to consider include:

- out-of-hours technical and experiment support
- post-experiment aftercare and bespoke analysis software
- laboratory space and subsidiary facilities for sample preparation
- food and scenery (you may be spending a couple of weeks performing an experiment!)

4.4   What Type of Proposal Should You Submit

Most facilities offer variations on the following access routes.

- Normal proposal submission route (aka Direct Access).
  - Proposals are submitted following calls for access applications. There are typically two calls per year. All Direct Access proposals are peer reviewed. Proposals that are allocated beam time are scheduled normally between 6 and 9 months after acceptance.

- Access for urgent studies (aka Rapid Access or Director's Discretion).
  - There are a variety of reasons why Rapid Access might be needed: a new material with high societal impact has been discovered; samples have short lifetimes; PhD student or post-doc needs beam time before their project ends *etc.* Rapid Access beam time proposals can be submitted at any time during the year and will be peer reviewed within a couple of weeks of receipt. Proposals allocated time will be scheduled as soon as possible. It is *essential* to discuss a Rapid Access proposal with facility staff before submission to ensure that beam time is available and that the requirements of the experiment are not so complicated that they prevent a rapid study.

- Access for feasibility studies (aka Express Access or Easy Access).
  - This route is ideal for straightforward, short measurements that are run on your behalf by facility staff. For example, a single measurement may be needed to demonstrate that a sample is suitable for a QENS study; the result being used to underpin a more developed Direct Access submission. Express measurements need to be simple, with no complicated sample environment or safety considerations. Samples are sent by courier to a facility, with reduced and corrected high-quality data returned ready for analysis, along with your sample.

- Commercial or industrial access.
  - Private sector researchers can use facilities *via* standard proposal routes in collaboration with an academic partner. However, dedicated schemes are also run for 'in-confidence' industrial or commercial access. Here, a fee will most likely be applied and it is essential to discuss a 'commercial' proposal with an Industrial Liaison Officer or relevant facility staff member.

- Programme access.
  - Facilities may be in a position to offer Programme Access support aimed to build, for example, science and innovation capacity in partner countries. These programmes encourage inter-country collaboration, which helps develop neutron research curricula in areas such as global challenge problems.

4.5 The Science Case

The heart of your application is the science case. This 2–3 page document should give a clear account of the aims of the experiment and be set within a broader scientific context. Keep in mind that review panel members might not understand the finer details of your research.

So, what does a successful science case contain?

First, a detailed, but concise, description of the problem, and its wider impact, is paramount. Explain why neutrons are needed and the reasoning behind your choice of instrument. Do not rely on the reviewer reading a library of referenced material. If you need to refer to a key finding in the literature then state it clearly in your text.

Second, where possible, summarise any preliminary work carried out by you and your team (for example, NMR or light scattering experiments) in support of your proposed experiment and to demonstrate sample quality.

Third, list the number of samples and sample environment conditions (for example, temperatures, pressures, *etc.*) required and state exactly the measurement(s) you want to perform. Break down the experiment in such a way that the review panel can assess the number of neutron beam days requested. Be sure to include any additional beam days needed for equipment set up, calibration and sample temperature equilibration. Can your experiment be performed on another instrument at the facility? If so, list alternative spectrometers.

Fourth, if you have financial support from a Research Council (or other source) describe how your proposal feeds into the funding body's wider aims. It is also worth indicating if your research has potential economic or societal impact.

Fifth, supply a list of publications arising from the work of you and your team. As well as providing supporting information, a good publication record assures the review panel that, if awarded time, your experiment will most likely be written up for a peer-reviewed journal.

Sixth, while PDF copy is becoming more common, proposals are still reduced in size and printed (black and white) in hardcopy for review panel members. As a result, use all of the available space and ensure all figure/graph legends and titles are legible (11 pt text being a suggested minimum).

Finally, if your proposal is a resubmission, you must clearly address the comments made by the review panel about your previous application. Furthermore, continuation proposals should be accompanied by an experimental report that summarises the results from preceding work.

4.6   The Proposal Review Process

Considering the Direct Access route, the proposal review process begins when a facility invites research teams to submit access applications. Be aware that access application submission deadlines are *real* and *enforced*!

Once submitted, your proposal will first pass through an internal technical review, to check feasibility, so ensure that all safety information requested is supplied. A technical review will not stop a proposal from being peer reviewed scientifically. However, concerns and recommendations will be passed to the scientific review committee.

The scientific review process requires that each submitted proposal first be classified in terms of either scientific subject area or neutron method depending upon the facility. Classification is managed by the Principle Investigator from a list of available options. The application is then assigned to the appropriate facility science review board (aka panel or college) for comment. Considering the application process used at the Institut Laue-Langevin, France, the science headings used to classify proposals include: Applied materials science; Theory; Nuclear and particle physics; Magnetic excitations; Crystallography; Structure and dynamics of disordered systems; Structure and dynamics of soft-condensed matter.

Each science review committee usually consists of facility representatives plus a dozen science experts, each proposal being allocated a minimum of two panel members for review. Depending upon the facility, review comments may be either simply collated electronically (the reviewers working from their home institution) or first collated electronically as mentioned but with panel members also convening for a two-day round table discussion. Regardless of the method used, the proposals are ranked by the review committee and, if successful, access is awarded in full or in part as appropriate. After a final adjustment (*e.g.* for national balance) letters are sent to the Principle Investigators to either confirm facility access or to provide feedback regarding the panel's decision not to award time. If awarded access then facility staff will contact the Principle Investigator to discuss the scheduling of your experiment.

Note: Neutron facilities operate twenty-four hours a day, seven days a week. When planning your schedule *you must consider carefully* the size of the research team that will attend the experiment.

4.7   Remembered...But for the Wrong Reasons!

Finally, it is worth remarking that while it is easy to generalise and make grammatical errors when drafting a proposal, subsequent proof reading before submission is paramount. This is especially important if written English is not your strength. While you might brighten their day, you don't want the review panel to comment on your work for the wrong reasons,

'I am overwhelmed by the feeling that I have spent longer reading this proposal than the author spent writing it!'

Institut Laue-Langevin review panel member.

5 The Measurement

In this chapter we will consider:

- Pre-experiment considerations.
- Instrument configurations.
- Preliminary measurements and checks.

5.1 Initial Considerations

Before your experiment begins key decisions need to be made. Careful consideration of each can greatly aid both measurement and analysis.

5.1.1 Sample Thickness

The first question to ask is:

How much sample do I need?

In truth, the amount is actually sought by asking: *how thick should my sample be?* More often than not sample thickness is adjusted such that multiple scattering (MS) effects can be largely ignored. None the less, thickness optimisation will allow you to gauge the amount (mass in grams) of sample needed for your experiment. We will consider multiple scattering in Chapter 6. However, suffice to say multiple scattering corrections are challenging. As a result, most researchers avoid such post measurement processing by limiting sample thickness such that just 10% of the incident beam is scattered (*yes, 90% is discarded!*).

A Practical Guide to Quasi-elastic Neutron Scattering
By Mark T. F. Telling
© Mark T. F. Telling 2020
Published by the Royal Society of Chemistry, www.rsc.org

If only 10% of all incident neutrons interact with a nucleus, then there will be a 1% chance that this 10% will collide a second time, and a subsequent 0.1% percent chance of a third impact, *etc.* At this level multiple scattering contributions are deemed negligible. Be aware, however, that effects from MS may still be observed and one should be mindful of this during analysis. For neutron spin echo experiments, sample transmission should be approximately 90% for those studies interested in the incoherent scattering of neutrons, else 50% or greater.

To optimise thickness, the Beer–Lambert form is used, which states that the flux transmitted through a material is given by,

$$I_{trans} = I_{incident}\, \exp\left(-n\,\sigma_T t\right) \tag{5.1}$$

where n = number density (atoms cm^{-3}), σ_T = total scattering cross section/formula unit (barns, 10^{-24} cm^2) and t = sample thickness (mm). Remember that for a material that scatters 10% of the incident beam then $\dfrac{I_{trans}}{I_{incident}} = 0.9$. For neutrons, the product $n\sigma_T$ is termed the *attenuation coefficient* (sometimes denoted by μ) and commonly given in units of cm^{-1}. The number density itself can be calculated using $n = \rho_m N_A/M$ where ρ_m is the mass density (g cm^{-3}), N_A is Avogadro's constant (6.022 $\times 10^{23}$ atoms $mole^{-1}$) and M is the molar mass.

By way of a simple example, and ignoring any wavelength dependent absorption in the sample, you can calculate that a poly-ethylene $(C_2H_4)_n$ sample needs to be just t = 0.15 mm thick to scatter 10% of the incident beam.

What happens, however, if your sample contains an element with a dominant absorption cross section, σ_a, which cannot be replaced by a less absorbing isotope?

In this case it is not simple to define an optimal thickness since, for a 10% scatterer, the sample may become too thick for the neutrons to escape. Instead, we aim to maximise the fraction of neutrons scattered. If we assume a uniform shape to the sample then all neutrons travel within the material with approximately the same path length. In this case, optimal thickness is calculated using,

$$t = \frac{\ln\left(\sigma_a + \sigma_T\right) - \ln\left(\sigma_a\right)}{n\sigma_T} \tag{5.2}$$

As an example, consider manganese. This metal has a moderately large absorption cross section: $\sigma_a = 13.3 \times 10^{-24}$ cm^2 at $\lambda_i = 1.8$ Å. Suppose we are doing an experiment on a spectrometer with $\lambda_i = 5.1$ Å. What sample thickness is needed to maximise the scattered fraction? Remembering that the absorption cross section is wavelength dependent then, for $\lambda_i = 5.1$ Å, $\sigma_a = 13.3$ barns $\times$ (5.1 Å/1.8 Å) = 37.7 barns, $\sigma_T = 2.15$ barns, $n = 79 \times 10^{21}$ atoms cm^{-3} and consequently the most neutrons are scattered when $t = 3.2$ mm. Knowing t, the scattered fraction, Σ, can be determined using,

$$\Sigma = \exp\left(-n\,\sigma_a t\right) - \exp\left(-n\left(\sigma_a + \sigma_T\right)t\right) \tag{5.3}$$

with sample transmission, T, being determined using,

$$T = \Sigma / \left(1 - \exp\left(-n\,\sigma_T t\right)\right) \tag{5.4}$$

For our manganese example, $\Sigma \sim 2\%$ and $T \sim 38\%$, respectively.

5.1.2 Sample Container and Orientation

Most QENS experiments are performed using aluminium (Al) sample containers. Not only is this metal cheap and easily machined, but it is also largely transparent to neutrons (*i.e.* weak coherent, incoherent and absorption scattering cross sections). Of course, while incoherent scattering from such a cell type may be minimal, one should be mindful that on instruments operating at elevated Q, structural features in the form of Bragg reflections (coherent scattering) may be visible, as shown in Appendix 3.

Aluminium containers are, however, of little use for highly corrosive substances. Here, consider anodised, polytetrafluoroethylene (PTFE) coated or quartz sample containers instead. In addition, using untreated aluminium containers for aqueous solution studies will result in the slow accumulation of aluminium oxide. Such build up may eventually compromise your experiment, especially for those solution experiments where measuring times are of the order of days.

If an aluminium container is the best option then the next question to ask is: *which container geometry should I use?* Two options are available: flat plate and annular cell.

Annular sample containers (Figure 5.1(i)) are constructed such that the sample sits in the narrow annular space between two inter-penetrating thin-walled, hollow aluminium cylinders of differing diameter. By varying the diameter of the inner cylinder, different annular depths, thus sample thicknesses, can be appropriated.

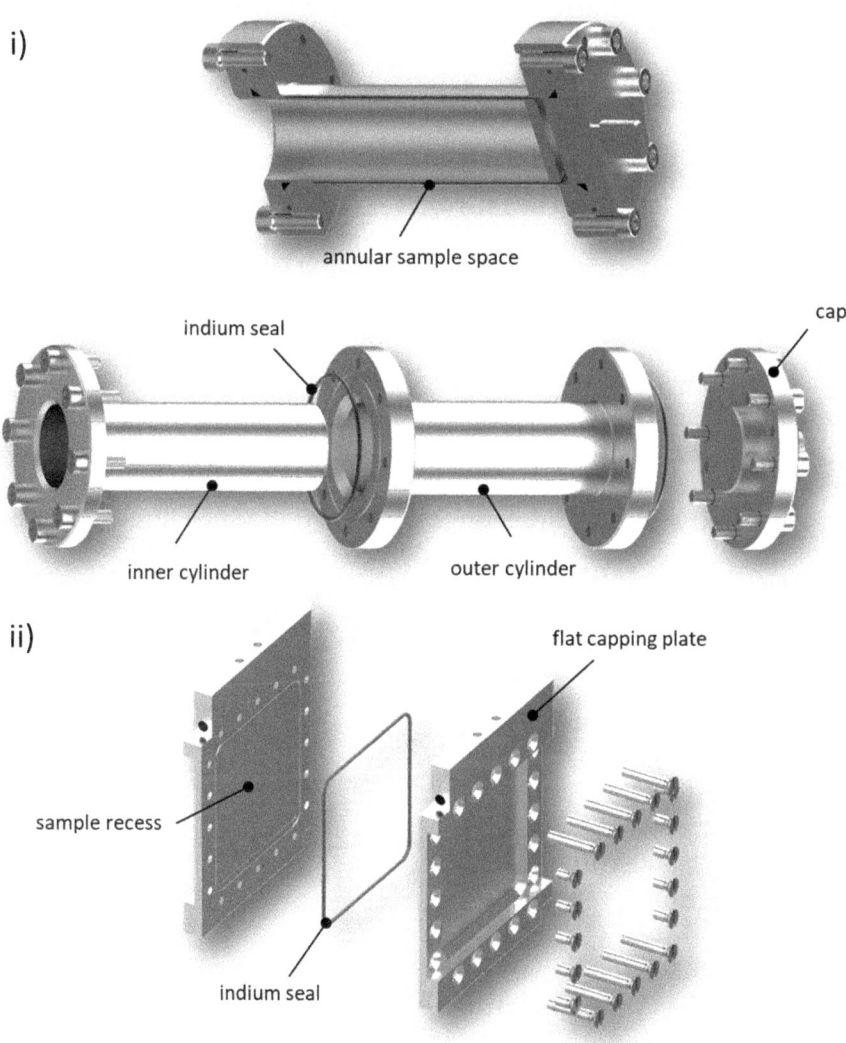

Figure 5.1 Illustrative examples of standard annular (i) and flat plate (ii) aluminium sample container assemblies. Designs will vary between neutron facilities.

Annular sample containers are ideal for:

- very fine powders
- liquids
- malleable thin films.

Scattering from an annular container is largely isotropic, with sample thickness being broadly constant as a function of scattering angle. However, one might notice subtle geometric effects (*i.e.* slight variation in detected intensity) as a function of Q.

For flat plate (Figure 5.1(ii)) sample cans, the sample sits in a recess machined into the surface of a thin Al slab. The recess has a height h, width w and depth d. In most cases, the height and width match the incident neutron beam cross section. However, the recess depth, d, is variable and should be chosen to match the sample thickness as determined using the Beer–Lambert equation. The sample is usually enclosed, and vacuum sealed, by placing a thin, flat Al capping plate (with indium wire seal) over the recess and fixing it securely into position using Al screws.

Flat plate containers are typically oriented in the neutron beam such that the assembly is set at either 45° (oriented towards low Q) or 135° (high Q) relative to the incident beam direction. Such alignment may result in a noticeable reduction in detected neutron intensity in those detectors viewing the container's edge. While corrections for such 'self-shielding' can be applied, the data collected in these 'masked' counters should be analysed with care.

Flat plate containers are better for:

- coarse/fine powders
- thin solids
- highly viscous materials
- non-malleable thin films
- small samples.

When using flat plate or annular sample containers it is worth considering whether your sample might settle at, or flow towards, the bottom of the container over time. Such movement may alter the amount of material in the beam, and thus explain unexpected variations in detected neutron intensity during your experiment.

Regardless of the cell type chosen, it is worth noting the following parameters while preparing your sample for subsequent data reduction and correction:

- the mass of the sample
- the mass of the assembled sample container before/after the experiment
- the thickness of the flat plate/annular aluminium walls in front of, and behind, the sample
- the recess depth, or annular width
- the inner and outer radii of the cylinders used to construct an annular container
- sample and container composition.

For those interested in high pressure measurements, thicker-walled cylindrical TiZr, Al, Ti–6Al–4V or BeCu cells are available, each having its own operating pressure and temperature range. For reference, key background features from the following empty container types are shown in Appendix 3.

- Aluminium (pressure cell)
- TiZr (pressure cell)
- Quartz (ambient pressure)
- Ti–6Al–4V (pressure cell)
- BeCu (pressure cell)
- Aluminium annular cell (ambient pressure).

Finally, no matter which cell type is used, it is wise to minimise background scatter by shielding all exposed wires, screws and container surfaces not directly in the neutron beam. This should be done using a neutron adsorbing material. The most common shielding materials are cadmium sheet (malleable and used for experiments where $T_{sample} < 373$ K) or gadolinium foil (less pliable but can be folded, or spot welded, to shape. $T_{sample} < 1000$ K). Bespoke boron based ceramic shields, *e.g.* B_4C, can also be machined for sample containers and apparatus designed to operate at elevated temperatures ($T_{sample} < 2500$ K).

5.2 Instrument Configuration

QENS instruments offer standard operating configurations that are suitable for most research needs (see Section 3.4). However, subtle changes to the mode of operation can be made that might benefit data interpretation and/or optimise data collection.

5.2.1 Symmetric vs. Non-symmetric S($\boldsymbol{Q}$,ω)

When working in Q–ω space it would be advantageous to record quasi-elastic scattering events over an infinitely wide energy transfer range ($\Delta\hbar\omega = +/- \infty$) in order to track broad spectral contributions, accurately determine EISFs and to gauge background intensity. In practice, however, no instrument affords such capacity. Instead, $\Delta\hbar\omega$ is finite and dependent upon instrument configuration. As a result, the information extracted may be limited due to inadequate extension in the wings of $S(\boldsymbol{Q},\omega)$; a broad spectral contribution, with a FWHM comparable to the instrument's energy transfer window, perhaps appearing as a flat background.

Most standard instrument set-ups are configured to measure symmetric data sets (*i.e.* $\Delta\hbar\omega$ straddles the elastic line equally in both positive and negative energy transfer). However, it can be advantageous to reconfigure the instrument to explore an asymmetric energy transfer response instead, especially for those systems in which two quasi-elastic signals are expected. An asymmetric measurement has key advantages. First, broad components can be followed to higher $\Delta\hbar\omega$ and thus modelled with greater accuracy, background levels being better defined. Second, while the measurement itself is asymmetric, line width analysis can be performed on symmetrised, or mirrored, data. Of course, care must be taken to apply the same mirroring to the associated instrument resolution measurement during analysis. It should also be mentioned, for those who prefer to transform $S(\boldsymbol{Q},\omega)$ and work in the time domain, that the wider the $\Delta\hbar\omega$ window probed, the better the definition of $I(\boldsymbol{Q},t)$ at short times. Such definition is important, especially where relaxation is fastest. It is also of consequence when deciding how best to interpret a normalised intermediate scattering function. For example, does the scattering function start at unity at zero time or is there suggestion of an additional rapidly relaxing component outside the observable time window of the spectrometer used, *i.e.* $I(Q,t = 0)$ appears to fall below 1.

Figure 5.2 illustrates narrow-symmetric and wide-asymmetric data, as well as a wide-symmetric $S_i(\boldsymbol{Q},\omega)$ response constructed by mirroring the asymmetric data measured between 0 and +1.0 meV. Both experimental data sets were collected from polydimethylsiloxane at 290 K. The inset illustrates how $I_s(\boldsymbol{Q},t)$ is better defined at short times when $S_i(\boldsymbol{Q},\omega)$ data collected using a wide energy transfer window is transformed.

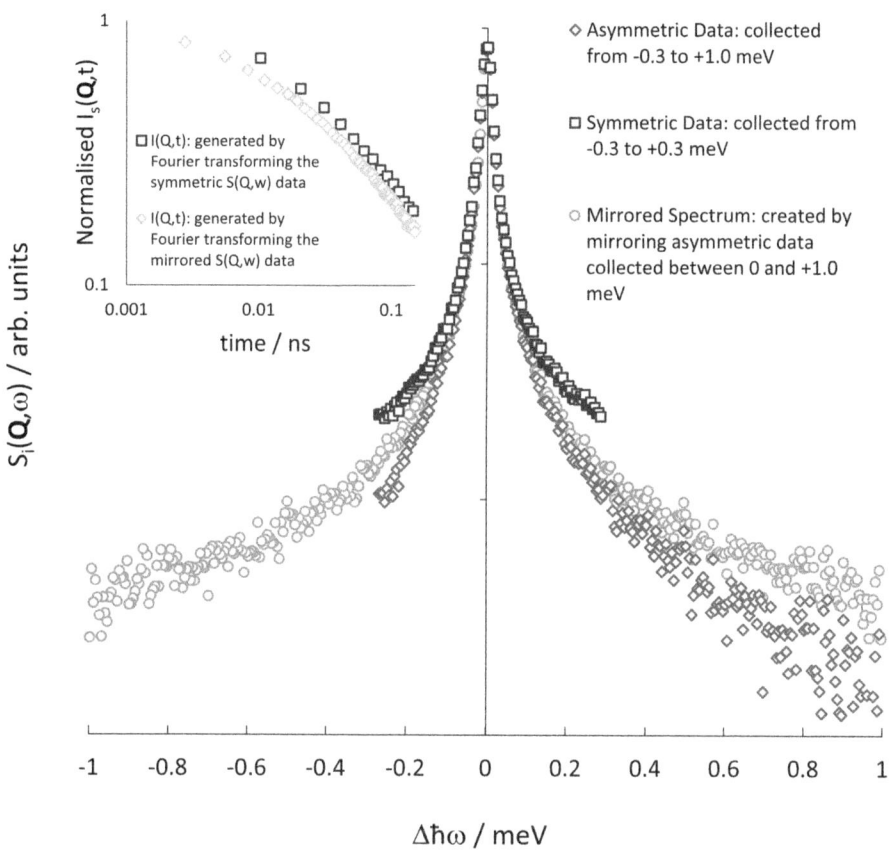

Figure 5.2 Measured narrow-symmetric and wide-asymmetric data sets alongside a wide-symmetric $S_i(\mathbf{Q},\omega)$ response constructed by mirroring the asymmetric data. Inset: the influence of $\Delta\hbar\omega$ on $I_s(\mathbf{Q},t)$ at short times. The data sets have been offset vertically to accentuate the differences between them.

5.2.2 Elastic Fixed Window Scan (EFWS)

The *elastic fixed window scan* (EFWS, $\mathbf{k}_i = \mathbf{k}_f$, $\Delta\hbar\omega = 0$) is a measurement method that involves observing the temperature dependence of just those neutrons scattered elastically as a function of Q, *i.e.* $S(Q, \omega \sim 0, T)$. While complementary to $S(\mathbf{Q},\omega)$ line width analysis, the popularity of the EFWS method stems from its fast data collection rate that affords a quick overview of any dynamical phenomenon over a wide external parameter range; typically, temperature. Indeed, EFWS measurements allow rapid access to transition temperatures (*i.e.* T_g, T_m), molecular rigidity ($\langle u^2(T) \rangle$) and Debye–Waller information ($d\langle u^2 \rangle/dT$). However more detailed analysis can yield, as an example, activation energies (E_a), mobile fractions (p_m) and associated geometry of motion ($A_0(Q)$).

The EFWS method is associated with direct or indirect geometry instruments, with information about the elastic scattering process alone being determined by noting only those neutrons scattered within a narrow energy interval about the elastic line. For instruments that operate in time-of-flight mode, the elastic scattering intensity is typically determined by integrating $S(\boldsymbol{Q},\omega)$ between $\Delta\hbar\omega_{min} = -\Gamma_{res}/2$ and $\Delta\hbar\omega_{max} = +\Gamma_{res}/2$. Here the width of Γ_{res} is defined at FWHM. For Doppler instruments, however, the elastic signal is recorded by measuring with the Doppler at rest. Examples of EFWS data, and subsequent analysis protocols, are given in Chapter 7.

5.2.3 Inelastic Fixed Window Scan (IFWS)

IFWS measurements are analogous to EFWS scans except that a finite energy offset, $\hbar\omega_{off}$, is probed ($\boldsymbol{k}_f$ fixed, $\boldsymbol{k}_i \neq \boldsymbol{k}_f$, $\Delta\hbar\omega \neq 0$). The method is associated with the Doppler-type instrument and the IFWS measurement is achieved by choosing a uniform, rather than sinusoidal, Doppler drive velocity profile.[1] Analysis of IFWS profiles, when used in combination with EFWS measurements, can help unravel complex dynamical information since the IFWS probes quasi-elastic scattering directly. Both IFWS and EFWS measurements allow access to similar experimental parameters. However, advantages of the IFWS method include (i) immediate identification of localised *vs.* translatory modes of diffusion; (ii) clear indication as to when the relaxation rate probed moves outside the experimental time window; (iii) clean separation of elastic and quasi-elastic signals and (iv) indication of transition temperatures and anharmonic behaviours. Examples of IFWS measurements, and subsequent data analysis protocols, are given in Chapter 7.

5.3 Preliminary Measurements and Checks

5.3.1 Instrument Resolution and Empty Sample Container

Any experimentally determined dynamic structure factor or intermediate scattering function is a convolution, or product, of a sample's true scattering response and the resolution of the neutron instrument used. Subsequent data analysis therefore requires an instrument's resolution function to be accurately described.

For instruments operating in Q–ω space, the instrument resolution approximates to, at its simplest, a Gaussian or Lorentzian function of finite width, Γ_{res}. The width of the resolution function affects an instrument's experimental observation time such that any observed quasi-elastic broadening will be associated with motions faster than $\tau^{-1} \cong \Gamma_{res}/\hbar$. Slower dynamics will be resolution limited (*i.e.* 'hidden' within the resolution function), while very fast processes will result in an extremely broad, background-like, component. Ideally, several energy resolutions should be used to avoid a resolution-biased view of any given problem. It is important, therefore, to accurately model the resolution response of any instrument you use, especially if your expected quasi-elastic scattering response is both narrow and lacks intensity.

For a well-defined resolution response, one might consider generating $R(\boldsymbol{Q},\omega)$ from its functional form. In practice, however, and especially for more asymmetric resolution functions, $R(\boldsymbol{Q},\omega)$ should always be measured during the experiment. In general, this measurement is performed using a standard; for example, a piece of vanadium similar in size, shape and orientation as the sample. However, $R(\boldsymbol{Q},\omega)$ can also be determined from the sample itself if the sample is cooled to a temperature where no pico-/nano-second motion is expected and, thus, only elastic scattering is recorded. The latter avoids uncertainties associated with a resolution function measured using a geometrically mis-matched standard. Be sure that cooling does not damage your sample.

For an NSE measurement, and as will be discussed in Section 6.3.3, the instrument resolution is removed from the measured intermediate scattering response by simply dividing each data point by $I(\boldsymbol{Q},t)$ collected from a wholly elastic scattering material.

Of course, to aid data reduction, a complete set of measurements should also include spectra collected from the empty sample container. Where applicable, this empty cell (aka background) measurement should be performed with the container oriented in the same manner as the sample. In practice, the detected scattering intensity from an empty container is usually negligible compared to the quasi-elastic response. Nonetheless, for completeness its contribution to $S(\boldsymbol{Q},\omega)$, or $I(\boldsymbol{Q},t)$, should always be considered. It should also be noted that an 'empty cell' measurement might not always be from an empty sample cell. For example, for those exploring adsorption of gas molecules on a surface, the background measurement might be from a sample cell plus substrate without gas; subtraction of the substrate/sample cell signal from the gas-loaded material allowing the research team to isolate the quasi-elastic scattering response alone.

5.3.2 Sample Transmission

While you should always calculate sample transmission before an experiment to optimise thickness and circumvent as best possible multiple scattering corrections, it is also worth determining T_{sample} experimentally. Not only does this approach provide a true transmission value for data reduction purposes, but for viscous or powder samples, for example, a transmission measurement performed before and after an experiment can help gauge whether your sample may have settled, flowed or leaked away from the neutron beam over time.

Most instruments include both incident beam (M_i) and transmitted beam (M_t) monitors, alongside bespoke software routines, to determine T_{sample} and any associated wavelength dependence. However, a simple check can be performed by noting neutron counts in said monitors at a particular incident neutron wavelength with and without the filled sample container in the beam. If R_{empty} is the ratio of the two monitor counts *without* the sample in the instrument (only cell and sample environment apparatus), and R_{sample} is the ratio of monitor counts *with* the sample in the instrument then the transmission, at the wavelength of choice, is $T_{sample} = R_{sample}/R_{empty}$.

5.3.3 Five Simple Rules

We conclude by suggesting guidelines that, if followed, will *(believe me!)* save beam time, optimise use of the neutron instrument and avoid frustration; especially in the small hours of the morning when you simply crave sleep. These checks should be performed before cooling or heating your sample. Therefore, before starting a measurement and leaving the instrument ask yourself:

- Is the shutter 'open'?
  - When changing a sample, access to the sample position and/ or any sample environment apparatus is *only* permitted if the spectrometer has been 'made safe', *i.e.* there is no chance of irradiation when working on, or around, the instrument. Isolating the spectrometer from the neutron source is achieved using a thick, neutron absorbing block (aka shutter). Button controlled, and mechanically driven into, and out of, the path of the neutron beam, *you* will be responsible for shutter control and research success relies on *you* opening it before starting to measure! To guide you, warning lights (usually red) set about the instrument should indicate that the shutter is open,

while elevated readings on radiation monitors will indicate that neutrons are entering the experimental area.

- Are neutrons being detected?
  - After a few minutes data collection, perform basic data reduction and visualisation to confirm that neutrons are reaching both the incident beam monitor and all detectors. This check will *also* confirm that the sample is in the neutron beam and that you are measuring a quasi-elastic scattering response.

- Is the sample orientation correct?
  - While not an issue for annular containers, flat geometry samples/cells are oriented such that the edge of the sample container points towards either high or low Q. The orientation chosen depends largely on the information to be extracted. Noting mis-alignment early is paramount, especially before significant amounts of data have been collected, or you start to cool, or heat, your sample. The simplest way to check orientation is to plot the elastic scattering intensity (*i.e.* neutron counts at $\Delta\hbar\omega = 0$) *vs.* θ and note the angle at which the recorded neutron intensity is a minimum.

- Is the heater connected and/or working?
  - Cooling a sample to base temperature using a cryostat or a closed-cycle refrigerator (CCR) takes at least one hour, possibly longer. Check, therefore, that heat can enter the system and warm your sample before starting to cool. Try raising the sample temperature by a couple of degrees to ensure the heater is operational.

- Have I checked the instrument control script?
  - Most neutron instruments are controlled using scripts. The scripts define which instrument configuration(s) are to be used and the sequence in which measurements are to be performed. Be sure to check the structure of your script carefully and note any syntax omissions and/or misspellings before launching.

Reference

1. B. Frick, J. Combet and L. van Eijck, *Nucl. Instrum. Methods Phys. Res., Sect. A*, 2012, **669**, 7–13.

6 Data Reduction

In this chapter we will consider:

- The basics of raw data reduction.
- Reducing the data: direct and indirect geometry instruments.
- Reducing the data: neutron spin echo.

6.1 Basics of Data Reduction

Before data analysis can commence, the raw data collected in each individual detector has to be 'reduced'. Data reduction involves applying a standard set of mathematical procedures to correct for, or remove, unwanted instrument, scattering and/or source contributions. In reality you don't need to understand the finer details of the reduction process since these are taken care of by well-established routines within the data reduction software package employed; the more familiar reduction and analysis packages in use today being the Large Array Manipulation Program (LAMP),[1] the Manipulation and ANalysis Toolkit for Instrument Data (Mantid)[2] and the Data Analysis and Visualisation Environment (DAVE).[3] While the package promoted may vary between neutron facilities, or even be bespoke code, the underlying principles should align. Indeed, the stages outlined below may be adopted, or overlooked, depending on the instrument and/or neutron source used or the level of detail to be extracted from the data. However, good practice suggests that each be considered.

A Practical Guide to Quasi-elastic Neutron Scattering
By Mark T. F. Telling
© Mark T. F. Telling 2020
Published by the Royal Society of Chemistry, www.rsc.org

6.2 Reducing the Data: Direct and Indirect Geometry

To begin we will consider the main data reduction procedures used to process data collected on an indirect or direct geometry instrument. We will then conclude by highlighting reduction protocols associated with data sets measured using neutron spin echo instrumentation.

6.2.1 Normalisation, Masking and Detector Efficiency Correction

At a minimum, the following corrections and considerations are required, with the three usually being performed in the sequence outlined below:

- accumulated incident beam intensity and/or wavelength profile normalisation
- neutron detector efficiency correction
- detector grouping or masking.

6.2.1.1 Incident Beam Normalisation

In order to draw comparison between measurements, each data set should be normalised to its associated accumulated incident beam intensity. For those instruments that operate using a band of incident energies, accumulated intensity may also show a wavelength dependence. For the purpose of normalisation, therefore, neutron instruments incorporate an incident beam monitor located upstream of the sample position. Some instruments also include a transmitted beam monitor placed downstream of the sample position from which sample transmission, and thus experimental sample scattering cross sections, can be determined. A typical incident beam monitor profile for a backscattering instrument on a pulsed source, and the effect of monitor normalisation, is illustrated in Figure 6.1 (i) and (ii).

6.2.1.2 Detector Efficiency Correction

Once the individual QENS spectra have been normalised to the accumulated incident beam intensity/wavelength profile, they are then corrected for detector efficiency. In practice, each detector

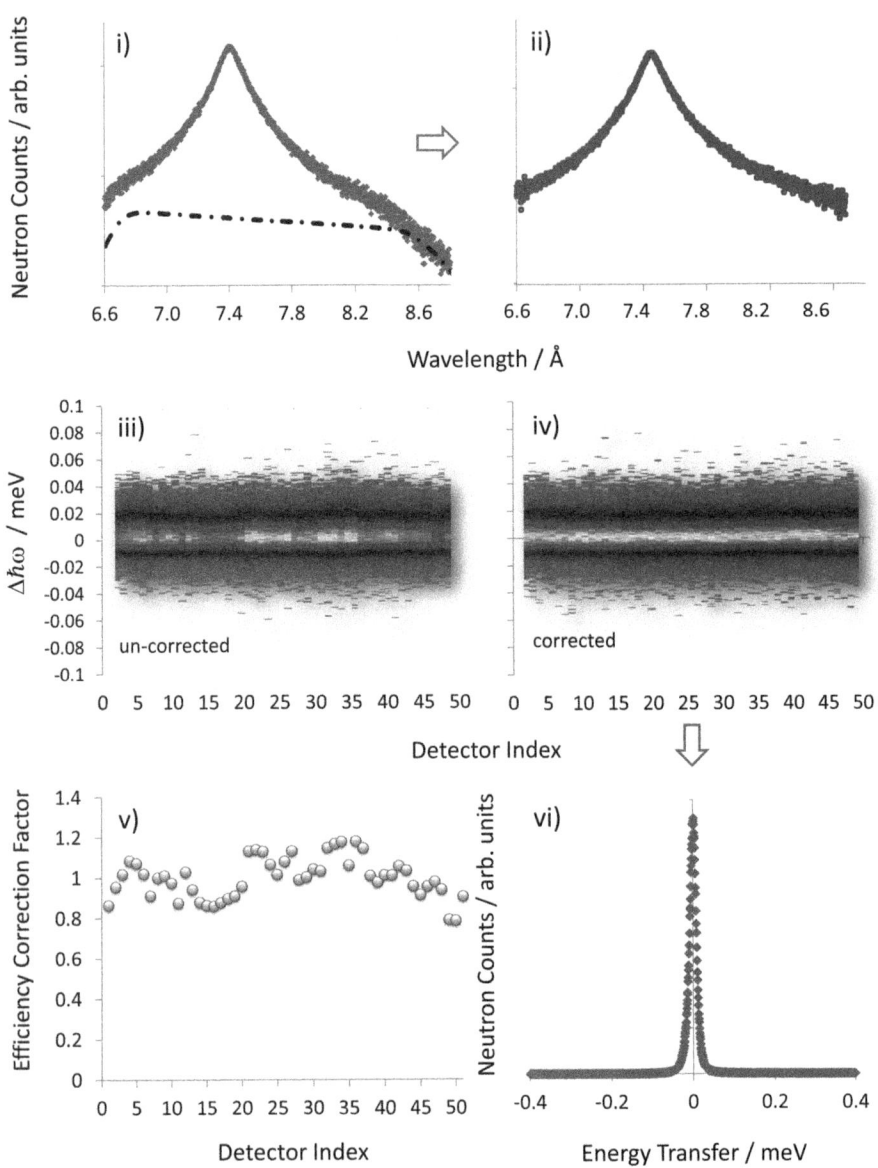

Figure 6.1   Top. Incident beam normalisation. (i) Broad peak: raw neutron counts in a single detector. Dashed line: incident neutron beam wavelength spread collected in the incident beam monitor. The rise and fall of incident beam intensity in (i) is due to the opening and closing of the chopper system used to define ΔE_i. (ii) The effect of normalising the raw data to the incident beam monitor profile. Middle/Bottom: The effect of detector efficiency correction. (iii) Plan view. Raw vanadium cylinder scattering data before correction. Note the variable elastic peak ($\Delta\hbar\omega = 0$) intensity between detectors due to unequal detection efficiencies. (iv) Detector efficiency corrected vanadium cylinder data. (v) Variation of the detector efficiency correction factor as a function of detector. (vi) Slice through (iv) to show the data in detector 25.

exhibits a slightly different detection efficiency. The result of such variation across a detector array is an observable fluctuation in detected neutron flux, especially noticeable at the elastic line. To correct for such variability, $S_i(Q,\omega)$ data is collected from a sample that scatters neutrons both elastically and isotropically (usually a thin-walled vanadium cylinder, ~10% scatterer) and from which a single value efficiency correction factor is generated for each detector (Figure 6.1(v)).

This instrument calibration measurement should be performed before your experiment begins, with the resulting detector efficiency factors being subsequently applied to all data sets during the data reduction process. If you want to ensure that detector efficiency corrections are being performed correctly, apply the efficiency factors to the data set from which they were generated. As seen in Figure 6.1(iv), the result should be a constant, scattering angle independent elastic peak intensity.

6.2.1.3 Grouping and Masking Detectors

Measurements are rarely of such superior quality that the data collected in each of an instrument's individual detectors can be analysed alone. Normally, spectra in neighbouring detectors are grouped and added during data reduction. While grouping helps improve statistics it does so at the expense of detailed Q, or length scale, information.

Note: when planning how best to group detectors, remember that Q does not vary linearly across a detector bank but follows $\sin(\theta/2)$; Q varying more slowly between detectors that extend high scattering angles relative to the straight through beam direction.

In contrast, masking (*i.e.* removal of individual detectors from the subsequent analysis process) may prove unavoidable should 'noisy' detectors, or Bragg features, contaminate a data set. Experimental data sets that exhibit (i) *extreme* Bragg contamination and (ii) 'dead' detectors, and for which masking is required, are illustrated in Figure 6.2.

6.2.2 Absorption Corrections and Empty Can Subtraction

When reducing scattering data, correction for intensity attenuation due to wavelength dependent neutron absorption in the sample and sample container (if present) should be considered. For a flat plate geometry such corrections can be analytical and are discussed, for example, by Carlile.[4] The situation for an annular container profile, however, is more complex and requires numerical integration.

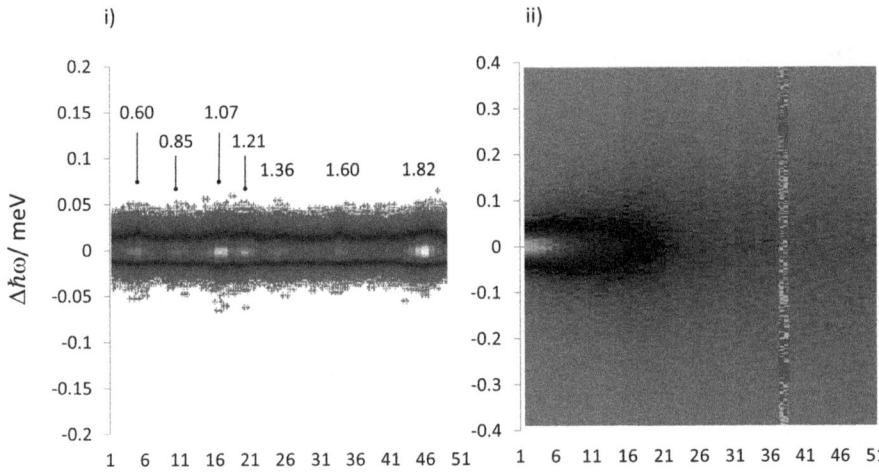

Figure 6.2 (i) Bragg contamination at the elastic line arising from the long-range crystallographic order in the sample (a metallic–organic framework (MOF)). The elastic line Q values associated with each Bragg peak are given. Clearly such extreme contamination limits the range over which detailed QENS information can be extracted with ease. A quick inspection suggests that detectors 28 to 34 (Q = 1.42–1.58 $Å^{-1}$) and 37 to 44 (Q = 1.65–1.77 $Å^{-1}$) can be used with care. (ii) 'Noisy' detectors, 38 and 39. These should be removed by instructing the data reduction software to ignore them during data processing. Data collected from H_2O at 280 K.

Nonetheless, such arithmetic techniques are well developed, used routinely in liquid and amorphous diffraction and are described by Soper *et al.*[5] and Paalman and Pings.[6]

The application of absorption corrections to experimental data is usually a two stage process. First, weighted and Q dependent correction files based on the user supplied sample and sample-container geometry/composition information are generated. Predicted Q dependent absorption correction factors for illustrative scattering geometries are shown in Figure 6.3. Second, these correction files are applied to the experimental data, with scatter from the empty sample cell being subtracted. However, this approach isn't without inherent limitation. The absorption correction process is based on the assumption that a theoretically flat, or cylindrical, construct accurately describes the scattering problem. Clearly such descriptions do not account for container deformation, non-uniform sample thickness, discrepancy in the angular

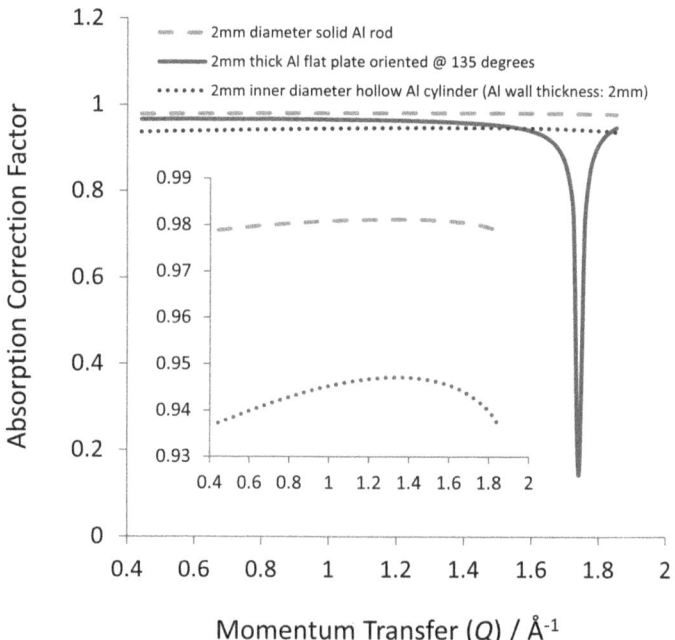

Figure 6.3   Absorption correction factors *vs.* momentum transfer as calculated using the Mantid package and the Paalman and Pings
method for (a) a 2 mm diameter, solid aluminium rod, (b) a 2
mm (inner) diameter hollow aluminium cylinder (wall thickness
= 2 mm) and (c) a 2 mm thick aluminium flat plate oriented
at a scattering angle of $\theta = 135°$ (relative to the incident beam
straight through direction). The loss of intensity at ~1.75 Å^{-1} for
the flat plate geometry is due to 'self-shielding' along the plate
edge.

orientation of a flat geometry sample, thermal expansion, contrac-
tion or flow. In practice, therefore, absorption corrected data needs
inspection before subsequent analysis. One observable to note is
over correction, which can result in an inverted peak at the elastic
line; a response most likely due to unrepresentative weighting and
thus subtraction of the empty container scattering contribution. In
most cases over adjustment can be compensated for by applying a
manual scaling factor.

6.2.3   An Alternative Philosophy

By following the procedures set out above, corrected data sets are
generated ready for conversion from $S(\theta,\omega)$ to $S(Q,\omega)$. However,
an alternative, commonly adopted, approach to data reduction
is to create the detector efficiency correction file from the sample

itself; $S_i(\mathbf{Q},\omega)$being measured to a high precision at a temperature low enough to suppress all motion(s) that occur within the observable time window of the neutron instrument used. By applying the resulting detector correction factors to subsequent data sets, this measurement-based philosophy offers an effective intensity scaling factor that not only accounts for detector efficiency deviations but, at the same time, scales experimentally, detected counts for neutron absorption. In addition, this method modifies directly any observed intensity variations arising from geometric effects, as opposed to relying upon the mathematical constructs used for the Paalman and Pings method.

In addition, considering subsequent line width analysis, this method provides a measure of a true instrument resolution profile as a function of scattering angle/Q from the sample *in situ*; $R(\mathbf{Q},\omega) = S(\mathbf{Q},\omega,\mathrm{T} = \text{low})$. Instrument resolution profiles can show subtle variation across a detector bank due to geometric effects, especially for flat plate geometries. As a result, the exact same resolution profile cannot always be assumed for each scattering angle. While such subtle difference has little effect for systems that show marked quasi-elastic broadening, it does have important consequences for accurate line width extraction from systems exhibiting weak quasi-elastic signatures that fall at the base of the elastic line.

It should be noted, however, that while an advantageous approach in many ways, this approach is not suitable for all systems. For example, such a correction philosophy should not be used for samples that show evidence of coherent scattering contributions across the detector bank.

6.2.4 Conversion to S(**Q**,ω)

As the individual Q–ω space detector trajectories shown in Figure 6.4 illustrate, each detector records intensity as a function of a common scattering angle, $S(\theta,\omega)$, not momentum transfer. As a result, Q is seen to vary in each detector as a function of energy transfer. While this variation may be considered negligible for small energy transfers, as one goes to larger $\Delta\hbar\omega$ the range of Q values spanned in a single detector can be sizeable and, potentially, skew analysis. As an example, the first detector in Figure 6.4 straddles a Q range from ~9.2 Å^{-1} (at $\Delta\hbar\omega$ ~ −65 meV) to 3.5 Å^{-1} (at $\Delta\hbar\omega$ ~ 20 meV). Good practice dictates, therefore, that data sets at constant momentum transfer be constructed from the raw data, *especially if one is to perform multiple scattering corrections*. To do so, one should use data reduction routines

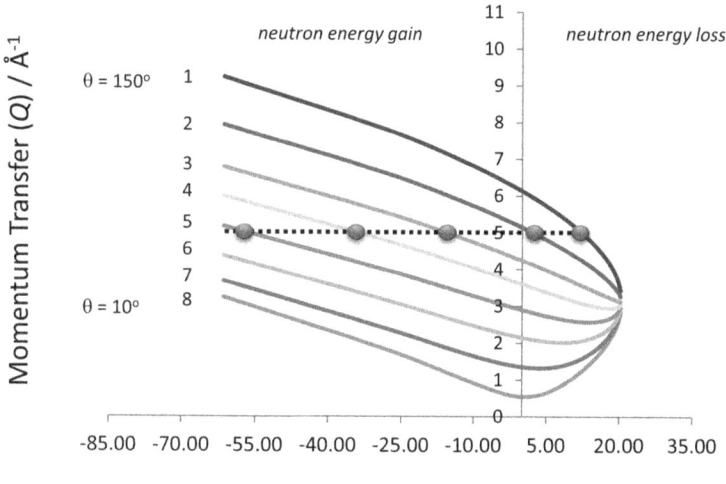

Figure 6.4 Illustrative Q–ω trajectories from eight detectors subtending scattering angles of 10° to 150°. Trajectories calculated using eqn (2.6), with E_i = 20.5 meV (~2 Å). As way of illustration, to construct an intensity *vs.* energy transfer spectrum at constant Q (= 5 Å^{-1}) one needs to interpolate between data from detectors 1–5 at the appropriate energy transfer values.

that cut through θ–ω space such that data is sampled from different detectors at constant Q. As one can appreciate, a certain amount of interpolation is needed between data points in order to construct a smooth constant Q image. As a result, high statistic measurements using instruments with high detector coverage, fine energy binning and tight Q resolution are of benefit here.

6.2.5 Multiple Scattering Considerations

Challenging, and subjective, multiple scattering (MS) corrections are usually 'overlooked' if sample transmission remains high (~90%) since, at this level, MS effects are considered negligible. As stated by Bée,[7] such an assumption is reasonable if the dynamical information being probed is unambiguous and restricted to the determination of physical parameters (*i.e.* the measure of characteristic times). However, data that is used to provide a detailed, quantitative comparison with such as the EISF, must be extracted to high precision if predictive models based on the assumption of single scatter are to be used to describe the experimental data. A basic understanding of the influence of MS on a measured data set is therefore beneficial.

Multiple scattering (MS) arises when the mean free path of the neutron inside a sample becomes such that there is elevated probability that a scattered neutron will interact with a second (and maybe a subsequent third) nucleus before exiting the sample. The mean free path is defined as the averaged distance between two successive absorption or scattering processes and is seen to be inversely proportional to the *total scattering cross section per unit volume*, Σ_T (= $\Sigma_a + \Sigma_s$). As one can appreciate, multiple scattering skews both the detected neutron's true 'single scatter' (aka first order) trajectory and energy transfer information, thus $S(Q,\omega)$. Indeed, even if a neutron does emit along a likely trajectory, it may have been 'back scattered' first, resulting in additional modification to its final energy (Figure 6.5). As a result the measured structure factor needs to be corrected for such alteration.

MS correction tools used nowadays are underpinned by the early Monte Carlo (MC) simulation packages developed by Bischoff (MSC[8]), Copley (MSCAT[9]) and Johnson (DISCUS[10]). Today, Monte Carlo protocols continue to dominate since they effectively simulate actual scattering experiments and different instrument geometries. However, the main advantage of today's MS correction suites is the availability of fast computing times for such computationally demanding calculations. Paradoxically, the MS correction itself ideally requires knowledge of the very structure factor, $S(Q,\omega)$, one is trying to determine experimentally. Since this is not available, calculations are, therefore, performed by first parameterising the experimental, non-corrected structure factor and monitoring the resulting $S(Q,\omega)$ responses after different orders of multiple scatter have been considered. The modified structure factor is compared with the experimental data and, providing a difference is observed, the input parameters to the MC evaluation are altered and the simulation re-run. Several iterations are usually required to converge on a solution.

For an expansive discussion on this subject you are referred to the work of Bée (Chapter 4[7]) with illustrative applications of, and considerations regarding, MS correction being reported by, amongst others, Zorn,[11] Busch,[12] Bée[13] and Wuttke.[14] It is not easy to attribute a general MS response to all systems. In broad terms, multiple scattering significantly affects incoherent quasi-elastic scattering spectra at lower Q. First-order incoherent scattering increases with Q. However, multiple scattering is essentially Q-independent. As a result, the contribution from multiple scattering is usually more significant at low

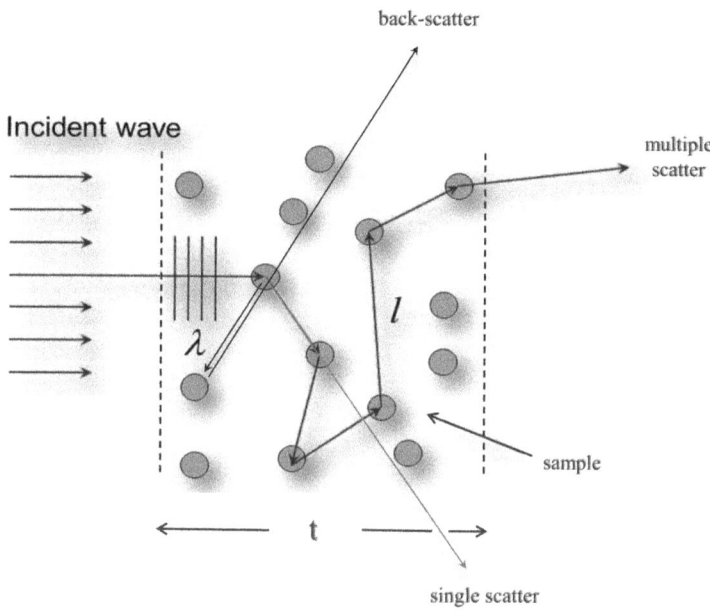

Figure 6.5 Single, back- and multiple scatter.

momentum transfer. However, for guidance, the influence of multiple scattering on your measured quasi-elastic scattering data sets might include:

- evidence of a multiple scattering component comparable in width to that expected from first order quasi-elastic scattering.
- evidence of a multiple scattering component that might be viewed as a flat background.
- under-estimation of the EISF at low Q (*i.e.* $A_0(Q = 0) \neq 1$) as shown in Figure 6.6. While a signature of MS, you should, however, check that deviation from unity at low Q is not due to coherent scattering, $S(Q)$, contamination.

6.3 Reducing the Data: Neutron Spin Echo

We conclude by considering the basic set-up, and data reduction considerations, associated with neutron spin echo instrumentation. It should be noted that this section deals with Mezei-type spin echo experiments. For the following discussion it is worth referring to Figure 3.6 in Chapter 3 and the accompanying instrument description.

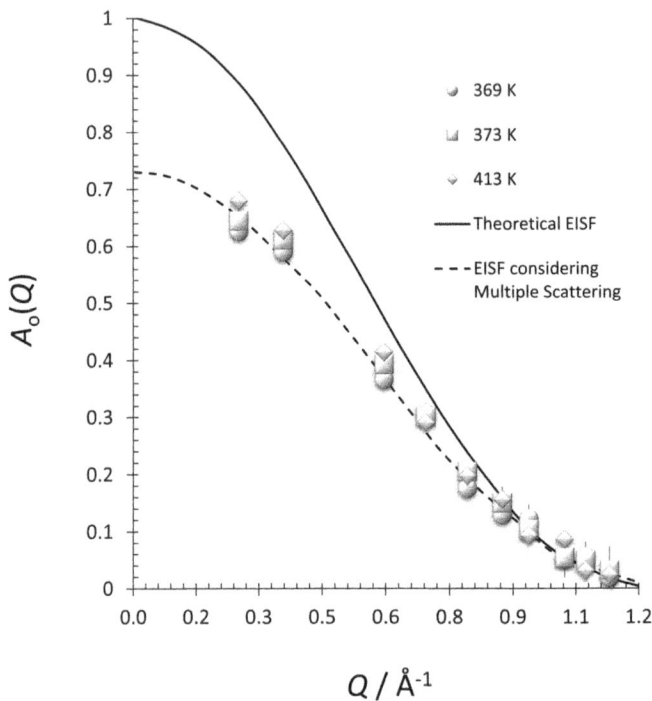

Figure 6.6 The effect of multiple scattering on the experimentally determined elastic incoherent structure factor of triethylene-diamine. The broken curve is the theoretical EISF with multiple scattering considered.[13]

At its simplest, an NSE instrument is configured by defining the incident wavelength band to be used and generating well-defined spin precession fields in the instrument's solenoids (aka coils or arms). Such parameterization defines the instrument's observable Fourier time, t_F, with neutron intensity at the detector being monitored in terms of the difference in field integral between the first and second coils. The wavelength spread itself is defined using a velocity selector to produce a band of finite width, $\Delta\lambda/\lambda_i$. This spread is between 10% and 20% at a reactor source but can be wider at a pulsed source. For single detector instruments, the momentum transfer vector of interest, and thus associated scattering angle, is also set.

6.3.1 The Echo

Initially, the NSE instrument is configured such that it satisfies the 'echo condition', *i.e.* the field integral in the first and second arms are equal. However, by varying the current flowing through one of the solenoids, controlled tuning/detuning of the initial neutron beam spin polarisation results. The coil in which current is varied is usually called the *phase coil*; the current flowing in the *phase coil* being related to the angle precessed by, and resulting phase (ϕ) of, the neutrons in the associated coil field. NSE data is therefore mapped in terms of detected neutron counts as a function of the phase for a specific Fourier time.

The variation of detected neutron intensity as a function of phase, ϕ, is illustrated in Figure 6.7. Such a response can be modelled using,

$$I = I_0 - A\cos[(\phi - \phi_0)/T]W(\phi - \phi_0, \sigma) \tag{6.1}$$

where,

- I_0 represents the average neutron intensity and is thus proportional to unpolarised static neutron counts.
- T is the period of the echo oscillation. The period is 360° within error. Sources of error include uncertainties in the average incoming wavelength value and the accuracy of the phase value when converting from coil current to precession angle.
- A is the amplitude at the echo point.
- ϕ_0 is the value of the phase at the echo point. While nominally zero, the echo condition might be reached for a non-zero value of the current in the *phase coil.*
- $W(\phi - \phi_0, \sigma)$ is the envelope function of the echo and is determined by the Fourier transform of the incoming wavelength distribution, $f(\lambda)$.
- σ is the width of the echo.

If $f(\lambda)$ is narrow, then the envelope, W, is broad. In the limit of a perfectly monochromatic beam, therefore, the echo would have no envelope and just be an infinite oscillatory function. In practice we assume that $f(\lambda)$ approximates to a Gaussian function such that,

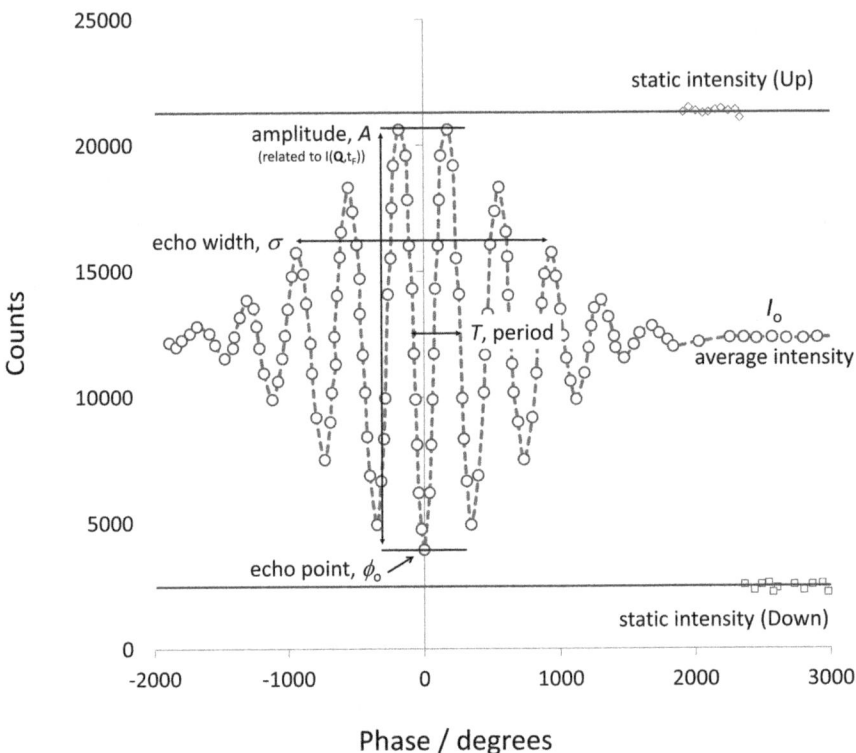

Figure 6.7   A typical echo. The main curve parameters are highlighted. The Up and Down measurements are artificially located at the end of the plot for the purpose of illustration.

$$W\left(\phi-\phi_{0},\sigma\right)=\exp\left[-\frac{\left(\phi-\phi_{0}\right)^{2}}{2\sigma^{2}}\right] \tag{6.2}$$

This assumption is general enough that it can be used in most cases. However, it is not difficult to modify the above equation to incorporate different incoming wavelength distributions. For example, and as used at the Spallation Neutron Source, ORNL, USA, if $f(\lambda)$ is rectangular then W is a *sinc* function.

The product of the oscillatory cosine function and the envelope W produces the typical echo shape seen in Figure 6.7. Information about the dynamics of the system is contained within the amplitude parameter, A. Since it is not possible to simply measure at $\phi = \phi_{0}$ with precision, measurements are carried out over a number of phase values. While the number of phase points measured may vary from one instrument to another, sampling always needs to be sufficiently large as to determine parameter, A, with satisfactory accuracy by fitting the data taken at each phase point to eqn (6.1). Such processing highlights a fundamental difference between NSE data reduction and that of other QENS-type measurements. For the latter, the reduction procedure relies on algebraic manipulation of the measured data. In contrast, for NSE measurements, least squared minimisation analysis is required.

Of the parameters in eqn (6.1), the period, T, and width, σ, of the echo envelope are instrumental parameters that only depend on the characteristics of the incoming wavelength distribution. As a result, they can be fixed. The phase of the echo point, ϕ_0, is also an instrumental parameter that does not depend on the sample. Nonetheless, it is supremely sensitive to the magnetic environment felt by the neutrons during their trajectory through the spectrometer and thus to changes in beam geometry, sample size, scattering angle and detection position. It is not, therefore, possible to determine this parameter with satisfactory precision before an experiment. On the other hand, since ϕ_0 does not depend on the sample itself, it can be determined experimentally using an appropriate standard; care being taken to ensure that the magnetic environment is stable.

Echo data collection is complemented by polarised beam diffraction measurements from which Up and Down spin static neutron intensities are ascertained. This measurement is performed by 'turning off' spin precession by not employing the $\pi/2$ flippers. Here, phase has no meaning, since there is no precession, and thus data can be taken with the *phase coil* switched off. This polarisation measurement is therefore performed with the π-flipper first on and then off to obtain the Up and Down counts, respectively.

Once the echoes collected for each Fourier time (Figure 6.8) have been fitted to determine A, a normalised intermediate scattering function (ISF) is obtained *via*,

$$\frac{I(Q,t_F)}{I(Q,0)} = \frac{2A(t_F)}{Up - Down} \tag{6.3}$$

The Up and Down counts are not dependent on t_F. Indeed, various strategies are possible to determine (Up-Down), which take advantage of the relation,

$$\frac{Up + Down}{2} = I_0 \tag{6.4}$$

In the case of experiments on paramagnets and diamagnets, the echo amplitude is normalised by one half of the magnetic scattering intensity as determined from full polarisation analysis.

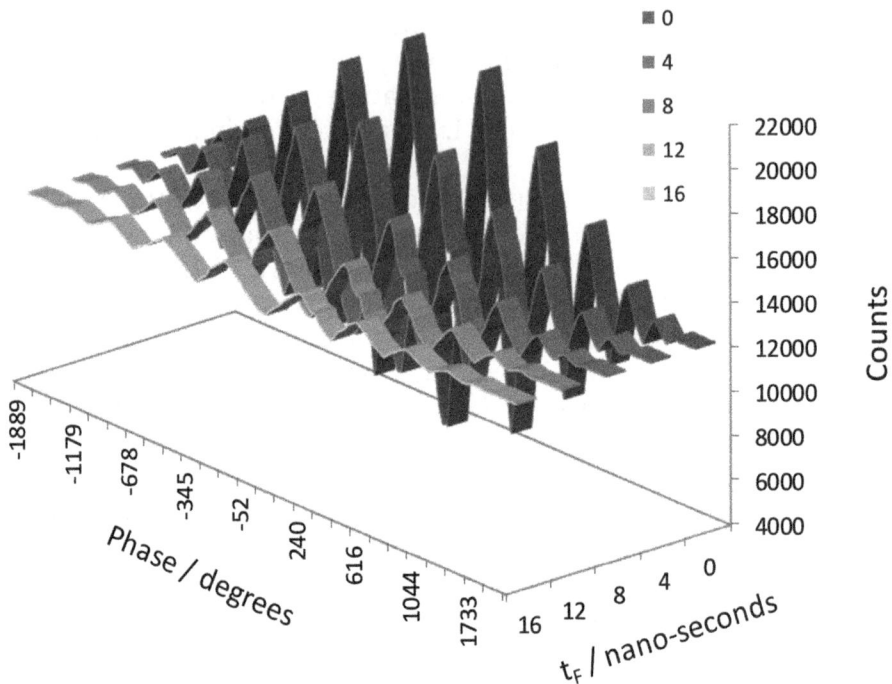

Figure 6.8 NSE data collected at various Fourier times as a function of phase. The reduction of the echo amplitude as a function of t_F is related to the dynamics within the sample.

Because the result of an NSE experiment is provided by the ratio between the echo amplitude and the polarised intensity, the total incoming beam flux, detector efficiency and instrument polarisation factors automatically cancel. Furthermore, while the incoming beam of an NSE spectrometer is generally only roughly monochromatic, it is assumed that the measured normalised intermediate scattering function is independent of the incoming neutron wavelength. Indeed, the incoming wavelength distribution is sufficiently symmetric and smooth that detailed wavelength dependent normalisation is not necessary. However, it should be kept in mind that the nominal Q and t_F values assigned to the intermediate scattering function correspond to those associated to the average incoming wavelength and an inherent spread is understood. This inherent spread is of the order of $\Delta\lambda/\lambda$ and $3 \times \Delta\lambda/\lambda$ for Q and t_F, respectively, and in most cases, not an issue since structural and dynamical features vary 'slowly'. The situation is slightly different on NSE spectrometers on a pulsed source since these instruments may operate using a rather large wavelength band but still retain the information due to narrow lambda bins. This situation does not modify the above consideration, rather it allows more flexibility, *a posteriori*, in the choice of $\Delta\lambda/\lambda$ for suitable Q and t_F accuracy.

6.3.2 Background Considerations

As with other quasi-elastic instruments, the need to subtract a background contribution might arise. For NSE, this operation should be carried out using the same considerations as for any other instrument. However, subtraction has to be performed independently for both the echo amplitude and the Up and Down measurements since the background may, and most of the time does, have a very different normalised intermediate scattering function compared to the sample. Consequently, before subtraction, the data needs to be normalised by the total incoming flux (or monitor counts, M).

The arguments presented in Section 6.2.2 generally apply for NSE. However, 'self-shielding' is more commonly taken into account *via* simple normalisation by the transmission. Furthermore, since NSE experiments are commonly carried out in solution, scattering from a solvent is usually also removed. The relevant equations here are,

$$\frac{I(Q,t)}{I(Q,0)} = \frac{I^{\text{exp}}(Q,t) - (1-V_s)I^{\text{bkg}}(Q,t)}{I^{\text{exp}}(Q,0) - (1-V_s)I^{\text{bkg}}(Q,0)}$$

$$= 2\frac{\dfrac{1}{M^{\text{exp}}T^{\text{exp}}}A^{\text{exp}} - (1-V_s)\dfrac{1}{M^{\text{bkg}}T^{\text{bkg}}}A^{\text{bkg}}}{\dfrac{1}{M^{\text{exp}}T^{\text{exp}}}\left(\text{Up}^{\text{exp}} - \text{Down}^{\text{exp}}\right) - (1-V_s)\dfrac{1}{M^{\text{bkg}}T^{\text{bkg}}}\left(\text{Up}^{\text{bkg}} - \text{Down}^{\text{bkg}}\right)}$$

$$\text{(6.5)}$$

where the suffixes *exp* and *bkg* refer to sample and background measurements, respectively. T is the beam transmission and V_s represents the volume fraction of scattering entities in solution. One effect that may be observed from a sample in solution is a normalised intermediate scattering function that does not extrapolate to unity at $t = 0$. Such divergence arises since the solvent exhibits fast dynamics outside the NSE time window. However, subtraction of the solvent contribution often restores the expected behaviour, as illustrated in Figure 6.9.

6.3.3 Instrument Resolution

A decrease in echo amplitude with Fourier time may not always be due to sample relaxation, but instead driven by instrumental effects. In particular, field integral inhomogeneities tend to reduce the echo

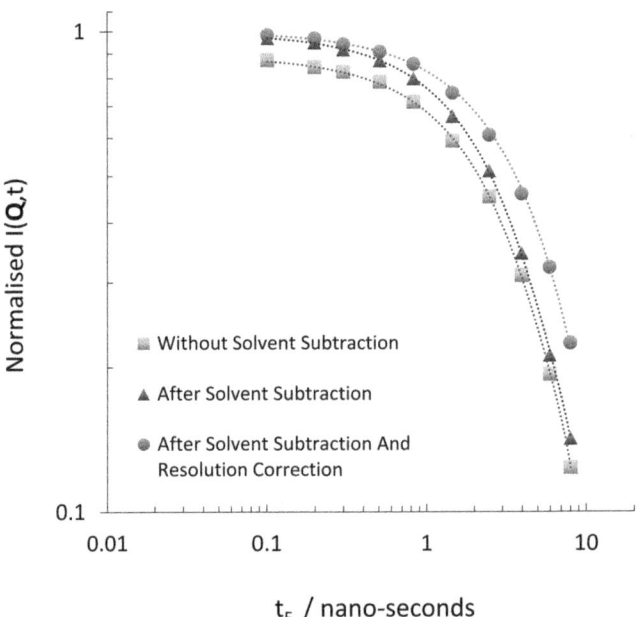

Figure 6.9   NSE measured normalised intermediate scattering functions from a micellar solution. The importance of solvent subtraction and resolution normalisation is highlighted.

amplitude as t_F increases. One can appreciate, therefore, that an echo measurement on a purely static sample would *still* result in a seemingly decaying normalised intermediate scattering function. This instrumental response from a purely elastic scattering sample defines the resolution function of an NSE instrument. Whereas for QENS spectrometers working in the energy domain the instrumental resolution cannot easily be removed from the data, on NSE, which works in the time domain, the instrumental resolution effect can simply be divided out,

$$\frac{I(\mathbf{Q},t)}{I(\mathbf{Q},0)} = \frac{\left[\dfrac{I^{\mathrm{exp}}(\mathbf{Q},t)-\left(1-V_s\right)I^{\mathrm{bkg}}(\mathbf{Q},t)}{I^{\mathrm{exp}}(\mathbf{Q},0)-\left(1-V_s\right)I^{\mathrm{bkg}}(\mathbf{Q},0)}\right]}{\left(\dfrac{I^{\mathrm{R}}(\mathbf{Q},t)}{I^{\mathrm{R}}(\mathbf{Q},0)}\right)}$$

$$= \frac{\left[2\dfrac{\dfrac{1}{M^{\mathrm{exp}}T^{\mathrm{exp}}}A^{\mathrm{exp}}-\left(1-V_s\right)\dfrac{1}{M^{\mathrm{bkg}}T^{\mathrm{bkg}}}A^{\mathrm{bkg}}}{\dfrac{1}{M^{\mathrm{exp}}T^{\mathrm{exp}}}\left(\mathrm{Up}^{\mathrm{exp}}-\mathrm{Down}^{\mathrm{exp}}\right)-\left(1-V_s\right)\dfrac{1}{M^{\mathrm{bkg}}T^{\mathrm{bkg}}}\left(\mathrm{Up}^{\mathrm{bkg}}-\mathrm{Down}^{\mathrm{bkg}}\right)}\right]}{\left(\dfrac{2A^{\mathrm{R}}(t)}{\mathrm{Up}^{\mathrm{R}}-\mathrm{Down}^{\mathrm{R}}}\right)}$$

(6.6)

Here, the suffix R refers to the instrument resolution measurement. The amplitude of the echoes for both the sample and the resolution are obtained from the fitting of the respective echo curves.

As long as there are no inherent dynamics on the timescale of the NSE method, the calibration standard used to determine the resolution response may vary. However, the material used should be shaped in a fashion similar to the sample under investigation, especially with respect to its size. If an NSE measurement is to be performed at low-Q, then carbon is most often used to determine instrument resolution because of its strong coherent scattering response in the low momentum transfer regime. For high Q, $Ti_{2.08}Zr$ is the most apt material due to its Q independent coherent scattering response. While vanadium may be used, it is not ideal given the fact that it is an incoherent scatterer and therefore the signal-to-noise ratio would reduce significantly. As discussed previously, the use of the sample itself to provide instrumental resolution data, cooled to a temperature low enough to suppress all relaxation, is an option that offers the advantage of ensuring that both sample and 'resolution' are from the same material and scattering geometry. The resolution measurement can also be used to determine, with good accuracy, the value of ϕ_0. If so, this parameter does not need to be fit during the data reduction of both sample and background measurements.

6.3.4 Detectors and Visualisation

Modern NSE spectrometers use large area, two-dimensional (2D) position sensitive detectors (PSDs). Of course, relatively speaking, they are significantly smaller compared to the m^2 size arrays used on a typical direct geometry instrument. For a standard general purpose NSE instrument, detector areas are of the order of ~20 cm × ~20 cm. Notable exceptions include IN11C (Institut Laue–Langevin, France), which had an angular coverage of 30° horizontally (1.5° vertically) and the WASP instrument (Institut Laue–Langevin, France) with its angular coverage of 150° horizontally (2.5° vertically). Of course, while a large detector increases the data acquisition rate, it is only of use providing that field integral inhomogeneities are maintained within acceptable limits across the detection surface.

In practice, echoes collected at different points on the detector may vary in terms of their echo point and resolution. Simply averaging the data, therefore, would result in a loss of information, if not full cancellation of the echo signal. As a result, NSE data collected across large position sensitive detectors are first binned in such a way as to improve statistics, but also to maintain the echo signal (Figure 6.10).

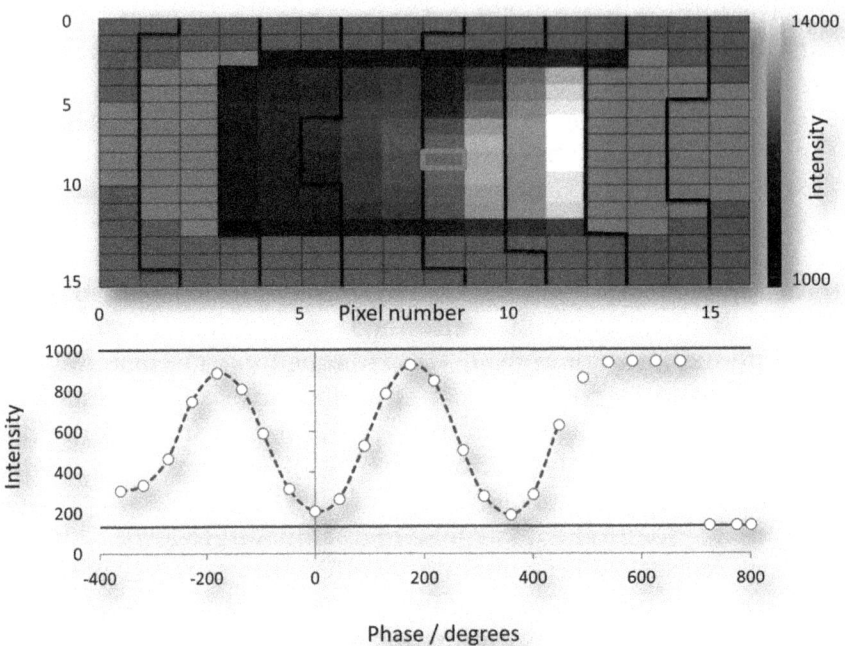

Figure 6.10   NSE reduction. Top: scattering intensity recorded on a single PSD on the CHRNS-NSE instrument at NIST, USA. The detector is 32 cm × 32 cm with 1 cm² resolution. However, the detector has already been binned 2 × 2 and some of the pixels masked. The thicker dashed lines between pixels delimit regions of equal Q. Bottom (right): the measured echo as contained in the highlighted pixel for a Fourier time of 0.05 ns. As can be seen, 19 phase points have been measured.

By limiting the bin size, aberrations in the binned data are minimised since each pixel data set is normalised by a broadly comparable resolution function.

Before binning, the data collected in each pixel on the detector is first independently reduced following the steps outlined above. Operatively, the resolution is usually analysed first to determine pixel dependent A^R and ϕ_0 parameters. Incidentally, the variation of ϕ_0 over the detector surface (from pixel to pixel) is commonly referred to as a *phase map* and should, ideally, be flat. The values of ϕ_0 are used in the fitting of both the sample and background echoes.

Once the normalised intermediate scattering function is determined per pixel, its value can be averaged over neighbouring segments of common Q. To ensure reasonable acceptance, an NSE detector can be divided into arcs at low scattering angles where Q variation is most significant. At high scattering angles, of course, such variation is less pronounced. Nonetheless, echo fitting has to be carried out independently for each pixel, even though, at high scattering angles, the whole detector could be averaged.

It is worth commenting that, as with other quasi-elastic scattering instruments, Bragg peaks should be removed from analysis. Since the incoming wavelength spread is large, Bragg peaks as seen using NSE tend not to be sharp but instead straddle a sizeable angular range. Contaminated data should be discarded if using a large angular acceptance detector. An exception to this is for NSE instrumentation on a pulsed source where the Q resolution can be finely tuned and studies can be performed at Q values close to a Bragg edge.

Since the reduction of NSE data involves complicated fitting procedures (often of a large number of data sets), the processing of NSE data is susceptible to error. Experience is invaluable for realising and quickly solving issues and, for those new to the method, guidance from facility staff is vital. However, as detailed above, the reduction of NSE data is largely exoteric.

Now we have addressed fundamental data reduction protocols, let's conclude by considering the analysis of our corrected $S(Q,\omega)$ or $I(Q,t)$ data sets. Part 3 therefore outlines key analysis strategies, model functions and probable responses. One can adopt one of two methodologies to parameterise the dynamic structure factor. While an initial examination of $S(Q,\omega)$ may involve the simple numerical integration of neutron counts, more involved least squares fitting, or Bayesian analysis, is usually required to extract quasi-elastic line widths and peak amplitudes *via* the convolution of Lorentzian functions and measured instrument resolution data. In contrast, transforming $S(Q,\omega)$ from the energy to the time domain serves to remove the instrument's resolution response leaving just $I(Q,t)$. While both analysis methods have merit, the fitting of the intermediate scattering function is usually the more straightforward, as we shall see in Chapter 8.

References

1. D. Richard, M. Ferrand and G. J. Kearley, *J. Neutron Res.*, 1996, **4**, 33–39.
2. O. Arnold, J. C. Bilheux, J. M. Borreguero, A. Buts, S. I. Campbell, L. Chapon, M. Doucet, N. Draper, R. F. Leal, M. A. Gigg, V. E. Lynch, A. Markvardsen, D. J. Mikkelson, R. L. Mikkelson, R. Miller, K. Palmen, P. Parker, G. Passos, T. G. Perring, P. F. Peterson, S. Ren, M. A. Reuter, A. T. Savici, J. W. Taylor, R. J. Taylor, R. Tolchenoy, W. Zhou and J. Zikoysky, *Nucl. Instrum. Methods Phys. Res., Sect. A*, 2014, **764**, 156–166.

3. R. T. Azuah, L. R. Kneller, Y. M. Qiu, P. L. W. Tregenna-Piggott, C. M. Brown, J. R. D. Copley and R. M. Dimeo, *J. Res. Natl. Inst. Stand. Technol.*, 2009, **114**, 341–358.
4. C. J. Carlile, *Rutherford Laboratory Report*, RL-74-103, 1974.
5. A. K. Soper, W. S. Howells and A. C. Hannon, *ATLAS – Analysis of Time of Flight Diffraction Data from Liquid and Amorphous Samples*, RAL Report RAL-89-046, 1989.
6. H. H. Paalman and C. J. Pings, *J. Appl. Phys.*, 1962, **33**, 2635.
7. M. Bée, *Quasi-elastic Neutron Scattering: Principles and Applications in Soild State Chemistry, Biology and Materials Science*, Adam Hilger, Bristol, England, 1988.
8. F. G. Bischoff, M. L. Yeater and W. E. Moore, *Nucl. Sci. Eng.*, 1972, **48**, 266–280.
9. J. R. D. Copley, *Comput. Phys. Commun.*, 1974, **7**, 289–317.
10. M. W. Johnson, *Discus: A computer program for the calculation of multiple scattering effects in inelastic neutron scattering experiments*, AERE-R-7682, United Kingdom, 1974, ISBN: 0705800547.
11. R. Zorn, *Nucl. Instrum. Methods Phys. Res., Sect. A*, 2007, **572**, 874–881.
12. S. Busch and T. Unruh, *J. Phys.: Condens. Matter*, 2011, **23**, 254205.
13. M. Bée, J. L. Sauvajol, A. Hédoux and J. P. Amoureux, *Mol. Phys.*, 1985, **55**, 637–652.
14. J. Wuttke, *Phys. Rev. E*, 2000, **62**, 6531–6539.

Part 3

Analysis

7 Elastic and Inelastic Fixed Window Scans

In this chapter we will consider:

- Analysis of elastic fixed window scan (EFWS) data.
- Analysis of inelastic fixed window scan (IFWS) data.

7.1 Elastic Fixed Window Scans (EFWS)

A wealth of information can be gleaned by considering just those neutrons scattered elastically. Indeed, for research teams new to the QENS method, analysis of elastic scattering is possibly the most straightforward and instructive. Information about the elastic scattering process alone can be determined by integrating neutron counts at each temperature typically between $\Delta\hbar\omega_{min} = -\Gamma_{res}/2$ and $\Delta\hbar\omega_{max} = +\Gamma_{res}/2$, rather than just logging the peak height at $\Delta\hbar\omega = 0$. Such integration is performed mostly to improve statistics. For Doppler instruments, the elastic signal is recorded by measuring with the Doppler at rest. The EFWS analysis method is associated with direct or indirect geometry instruments and analysis related to $S_i(\boldsymbol{Q},\omega)$. From here on in the elastic incoherent scattering intensity, $S_i(Q,\omega \sim 0,\mathrm{T})$, will be denoted $I_i^{el}(Q,T)$.

First, transition temperatures can be determined from EFWS analysis by simply plotting $I_i^{el}(Q)$ as a function of temperature, deviation from linear behaviour suggesting a change of state. Of course, the elastic scattering intensity will naturally decrease, albeit minimally, with

A Practical Guide to Quasi-elastic Neutron Scattering
By Mark T. F. Telling
© Mark T. F. Telling 2020
Published by the Royal Society of Chemistry, www.rsc.org

increasing temperature due to vibrations following the Debye–Waller form. However, as the frequency of a specific dynamic process within the sample (*e.g.* perhaps thermal activation of side-group motion in a soft matter system) enters the spectrometer's observable time window, a distinct inflexion, or marked reduction, in elastic information will be observed. A loss in elastic response that is not Debye-related will be accompanied by a gain in quasi-elastic scattering.

Example EFWS data is shown in Figure 7.1(i) for isotactic ($T_{g,\text{iPP}}$ = 261 K, $T_{m,\text{iPP}}$ = 436 K as determined using differential scanning calorimetry (DSC)), syndiotactic (DSC: $T_{g,\text{sPP}}$ = 269 K, $T_{m,\text{sPP}}$ = 403 K) and atactic (DSC: $T_{g,\text{aPP}}$ = 266 K) polypropylene.[1] The data have been reproduced from Arrighi *et al.*[1] for Q = 1.85 Å^{-1} where the EFWS method was used to investigate the influence of tacticity on the glass transition. The inflexion points in the EFWS data sets are consistent with thermal transitions as measured using DSC and above the glass transition temperatures the dynamical response is clearly tacticity dependent; the decrease in elastic intensity being due to the onset of segmental motion. The inflexion at T_{CH_3} (~100 K) indicates the onset of rotational motion associated with methyl groups below which the system is considered a harmonic solid. The Q and T dependence of the temperature normalised elastic scattering response, *i.e.* $\dfrac{I_i^{\text{el}}(Q,T)}{I_i^{\text{el}}(Q,T=10\text{K})}$, for isotactic polypropylene at Q = 0.5, 0.86, 1.45, 1.68, 1.85 and 1.96 Å^{-1} is shown in Figure 7.1(ii).

Second, sample 'rigidity' can be gauged by extracting the mean squared displacement (msd) parameter, $\langle u^2(T)\rangle$, from the EFWS data. This parameter corresponds to the average harmonic displacement amplitude of all atomic motions in the sample. If the elastic scattering intensity, at any given temperature, is monitored as a function of Q then an effective (all atom) $\langle u^2(T)\rangle$ value can be obtained by fitting,

$$I_i^{\text{el}}(Q,T) = \exp\left(-\frac{Q^2\langle u^3(T)\rangle}{3}\right) \qquad (7.1)$$

One should recognise eqn (7.1) as the Debye–Waller (DW) form. Plotting $\langle u^2(T)\rangle$ as a function of temperature, therefore, allows sample 'flexibility' to be appraised, with deviations from a purely Debye-like harmonic response at low temperatures signalling changes in structural integrity. Such an analysis approach is now used routinely in many areas of science.[2,3] It is worth commenting here that if the atomic motions probed using EFWS are localised, the msd will tend to a finite value at infinite time. In this case, the pre-factor used to

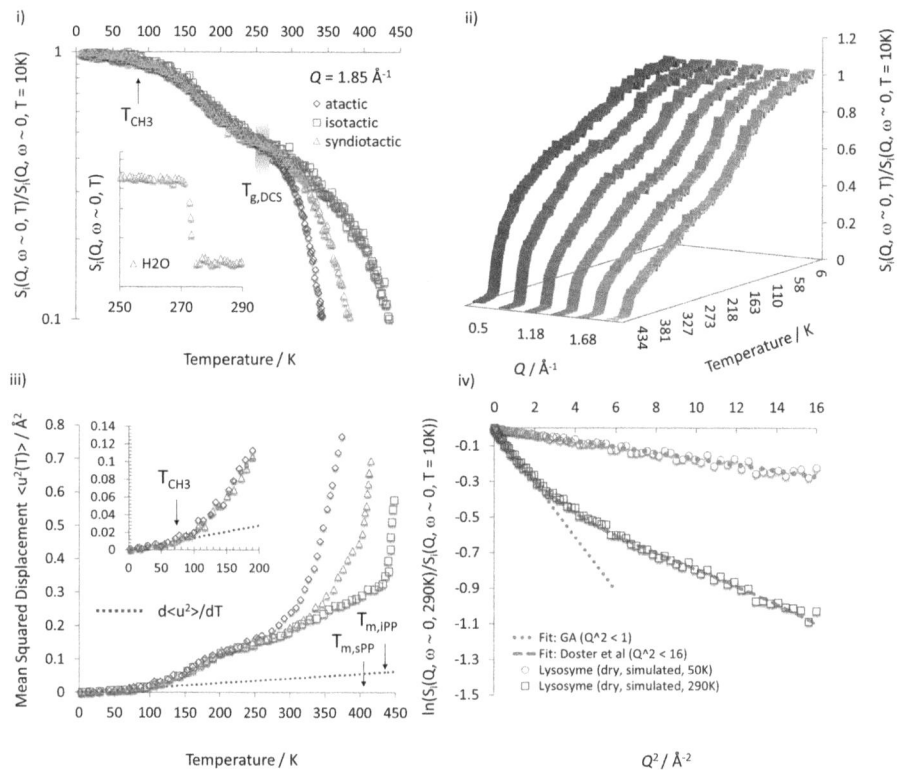

Figure 7.1 (i) EFWS data for isotactic, syndiotactic and atactic polypropyl-ene. Inset: for reference, the expected EFWS response from H_2O at the melting point. Data collected using IN13, Institut Laue–Lan-gevin, France. (ii) The Q and T dependence of normalised $I^{el}_i(Q,T)$ for isotactic polypropylene from which the $\langle u^2(T) \rangle$ parameters shown in (iii) were determined. (iii) Mean squared displace-ment parameters *vs.* temperature. Inset: rate constant, $d\langle u^2 \rangle/dT$, determination as discussed in Section 7.1.1. For polypropylene, a rate constant of $1.4(\pm 0.1) \times 10^{-4}$ Å^2 K^{-1} is reported.[1] (iv) Molecular dynamics simulated $I^{el}_i(Q,T)$ data for lyophilised lysozyme[5] at 50 K and 290 K. Eqn (7.2) provides an accurate description of the data at low temperature (50 K, fit range: $Q^2 = 0$–16 Å^{-2}) and for $Q^2 < 1$ Å^{-2} at elevated temperature. However, the onset of anhar-monic processes requires eqn (7.2) to be modified. Eqn (7.5), as way of an example, provides a more accurate description of the data over the entire simulated Q^2 range.

describe time independent (aka static) elastic scattering evaluates to 1/3. In contrast, for time dependent (aka time-resolved or dynamic) msd(t) measurements the pre-factor 1/6 is used.[4] Figure 7.1(iii) shows mean squared displacement parameters *vs.* temperature for isotactic, syndiotactic and atactic polypropylene as determined by fitting eqn (7.1) to their respective EFWS response.

The method of extracting $\langle u^2(T)\rangle$ itself exhibits potential limitations. First, by rearranging eqn (7.1) we see that,

$$\ln\left(I_i^{\text{el}}(Q,T)\right) = \left(-\frac{Q^2\langle u^2(T)\rangle}{3}\right) \tag{7.2}$$

Assuming purely harmonic motion, experimental data plotted in this fashion should exhibit a linear Q^2 dependence whose gradient is proportional to $\langle u^2(T)\rangle$. One can also appreciate that the y-axis intercept, $I_i^{\text{el}}(Q = 0, T)$, of the fitted data should be temperature independent. In practice, however, this is rarely the case. Poor statistics, limited momentum transfer information at low Q, coarse detector/Q coverage, possible coherent scattering contamination or multiple scattering effects (especially at small Q values) can skew the gradient.

Second, the efficacy of eqn (7.2) to fully describe the Q dependence of the elastic intensity is limited to situations where $Q^2\langle u^2(T)\rangle \ll 1$; the so-called Gaussian Approximation (GA) introduced by Rahman et al.[6] in 1962. Should anharmonic processes, such as heterogeneity of motion due to additional degrees of freedom, become evident, then eqn (7.2) no longer provides an accurate description of the data at all Q values and modifications are needed. As an example, simulated $I_i^{\text{el}}(Q,T)$ data for lyophilised lysosyme[5] at 50 K and 290 K is shown in Figure 7.1(iv). Eqn (7.2) provides an accurate description of the data over the whole Q range at low temperature (50 K, fit range: $Q^2 = 0\text{--}16$ Å^{-2}) but only for $Q^2 < 1$ Å^{-2} at elevated temperature. The onset of heterogeneity in the system at elevated T and Q requires eqn (7.2) to be modified.

It is worth mentioning here that of all areas of neutron science, it is the study of the bio-polymers, in particular hydrated proteins, that has led to the development of models that better describe $I_i^{\text{el}}(Q,T)$ beyond the GA. Overviews of such work can be found in Zeller et al.,[7] Doster[2] and Gabrielli[5] and references therein. In brief, these models account for heterogeneity by assuming harmonic motions of an individual atom and non-Gaussian behaviour arising from a distribution of individual, yet different msd parameters. Three models are summarised below.

The first, proposed by Peters and Kneller[8] (PK), uses a gamma function to describe the distribution of mean square position fluctuations (MSPFs). The PK model is not limited in Q. Here,

$$S_i(Q,\omega \sim 0,T) \approx \frac{1}{\left(1 + \dfrac{Q^2\langle u^2(T)\rangle_{\text{MSPF}}}{3\beta}\right)^\beta} \quad \text{and} \quad \sigma_{\text{MSPF}} = \frac{\langle u^2(T)\rangle_{\text{MSPF}}}{\sqrt{\beta}} \tag{7.3}$$

where $\langle u^2(T)\rangle_{MSPF}$ describes the mean of the individual MSPF and σ_{MSPF} is its standard deviation. Parameter, β, describes the distribution of the MSPF.

An alternative method of describing $I_i^{el}(Q,T)$ data beyond the GA was proposed by Yi *et al.*[9] as a modification to work introduced by Becker *et al.*[10] Here,

$$S_i(Q,\omega \sim 0,T) \approx \exp\left(-\frac{1}{3}Q^2\langle u^2(T)\rangle_{Yi}\right) \times \left(1+\frac{Q^4}{18}\sigma_{Yi}^2\right) \tag{7.4}$$

where σ_{Yi} describes the standard deviation of the mean square displacement. It should be noted that applicability of the Yi model is limited in Q.

Finally, Doster *et al.*[11] proposed a double potential well based model that assumes atoms can be found in one of two harmonic wells separated by a distance, d, and free energy difference, ΔG. The hydrogen atom is specifically considered in this work due to, as we have discussed, its sizeable incoherent scattering cross section and thus significant contribution to $S_i(Q,\omega)$. Here,

$$S_i(Q,\omega \sim 0,T) \approx \exp\left(-\frac{1}{3}Q^2\langle u^2(T)\rangle_{Do,G}\right) \times \left(1-2p_{12}\left(1-\mathrm{sinc}(Qd)\right)\right) \tag{7.5}$$

where the first term, $\langle u^2(T)\rangle_{Do,G}$, describes the Gaussian contribution to the msd and the second term describes the two state model. p_{12} denotes the probability of finding an atom in the ground (p_1) or excited state (p_2), with $p_2/p_1 \propto \exp(-\Delta G/RT)$.

It is also worth commenting that if the mean squared displacement parameter, $\langle u^2(T)\rangle$, is thought of as arising from a ball and spring like construct then an effective average force constant, $\langle k\rangle$, can be determined from which molecular 'stiffness' can be ascertained. By applying a quasi-harmonic approximation,[12] $\langle k\rangle$ can be calculated from the temperature dependence, or slope, of $\langle u^2(T)\rangle$ over specific temperature regimes using,

$$<k> = A\left(\frac{d\langle u^2\rangle}{dT}\right)^{-1} \tag{7.6}$$

The numerical constant, A, allows $\langle k\rangle$ to be expressed in Newton per metre when $\langle u^2(T)\rangle$ is given in Å^2 and T is the temperature in Kelvin.[13] As an example, D_2O hydrated myoglobin powder exhibits a force constant of $\langle k\rangle = 2\ \text{Nm}^{-1}$ below 200 K and 0.3 Nm^{-1} above.

EFWS analysis allows instructive information, and trends, to be extracted with minimal data processing. However, as we have seen for localised dynamics,

$$S_i(\mathbf{Q},\omega) = \exp\left(-\frac{Q^2\langle u^2(T)\rangle}{3}\right)\left(A_0(Q)\delta(\omega) + A_1 S_i^{qe}(\mathbf{Q},\omega)\right) \tag{7.7}$$

As stated, $A_0(Q)$ is the elastic incoherent structure factor (EISF) associated with spatially constrained motion and, at its simplest, the quasi-elastic scattering component is a single Lorentzian function whose width, Γ, represents the frequency of said mobility. Therefore, for temperatures above an observed inflection/transition, $S_i(\mathbf{Q},\omega)$, and thus $S_i(Q, \omega \sim 0, T)$, will comprise of both elastic *and* quasi-elastic information. While inclusion of a quasi-elastic contribution to the elastic line integral rarely affects broad systematic trends, accurate analysis of $I_i^{el}(Q, T)$ at elevated temperatures requires the quasi-elastic contribution to the elastic scattering intensity to be removed.

7.1.1 Basic Modelling of $I_i^{el}(Q,T)$

Following on from eqn (7.7), EFWS data itself can be analysed in terms of $A_0(Q)$ for localised motions. Grapengeter[14] showed that for a single type of mobile specie, moving at a defined frequency, the contribution of the quasi-elastic scattering component within a fixed energy window can be determined using,

$$I_i^{el}(Q,T) = \text{DWF} \times \left(A_0(Q) + \frac{2}{\pi}[1 - A_0(Q)]\arctan\left(\frac{\Gamma_{res}}{\Gamma}\right)\right) \tag{7.8}$$

Γ_{res} is the width of the spectrometer resolution function and Γ is the width of the Lorentzian line characterising the quasi-elastic broadening. In this model, it is assumed that the width of the quasi-elastic scattering component broadens according to the Arrhenius relationship,

$$\Gamma(T) = \Gamma_0 \exp\left(\frac{-E_a}{RT}\right) \tag{7.9}$$

where R (= 8.31 JK^{-1} mol^{-1}) is the gas constant. The activation energy, E_a, is related to the height of the potential barrier hindering localised motion while Γ_0 is the line width at infinite temperature. In contrast, should the specie type under investigation be subject to, say, a heterogeneous environment such that there exists a distribution of relaxation rates then,

$$I_i^{el}(Q,T) = \text{DWF} \times \left(A_0(Q) + \frac{2}{\pi}[1 - A_0(Q)] \times \sum g_i \arctan\left(\frac{\Gamma_{res}}{\Gamma_i}\right)\right) \tag{7.10}$$

where g_i gives the weight of each spectral component according to a Gaussian distribution of activation energies.

It should be mentioned that it is highly unlikely that all species in a material will contribute to the reduction of $I_i^{el}(Q,T)$. Assuming no coherent scattering contributions to the detected signal, arising from (for example) deuterated molecules, then the mobile fraction associated with the quasi-elastic response will depend upon both temperature and the spectrometer's experimental observation time. As a result a reduced, or effective, EISF parameter, $A_0'(Q)$, rather than the theoretical $A_0(Q)$, will most likely be measured.[15] Nonetheless, the true EISF can be deduced from this reduced parameter by noting that,

$$A_0'(Q) = p_f + (p_m \times (A_0(Q))) \text{ where } p_f + p_m = 1 \qquad (7.11)$$

Here, p_f and p_m are the relative proportions of fixed (*i.e.* static on the experimental timescale) and mobile atoms. As a result, eqn (7.10) might be written,

$$I_i^{el}(Q,T) = \text{DWF} \times \left(\begin{array}{c} 1 - p_m + p_m A_0(Q) + \dfrac{2}{\pi}\left[1 - \left[1 - p_m + p_m A_0(Q)\right]\right] \\ \\ \times \sum g_i \arctan\left(\dfrac{\Gamma_{res}}{\Gamma_i}\right) \end{array} \right) \qquad (7.12)$$

Similarly, for eqn (7.8). A detailed and more expansive number of predicted forms for $A_0(Q)$ are listed in Bée,[16] with more specific variations being derived in the literature. However, common responses encountered in a number of systems are given below and illustrated in Figure 7.2.

$A_{0,CH_3}(Q) = \dfrac{1}{3}\left(1 + 2j_0\left(Qr\sqrt{3}\right)\right)$	The EISF response predicted for a proton undergoing jumps between three equidistant sites on a circle of radius, $r(\text{Å})$. Descriptive of, for example, protons on a methyl group, CH_3. j_0 is the zero-order Bessel function: $j_0(x) = \sin(x)/(x)$.
$A_{0,jump}(Q) = \dfrac{1}{2}\left(1 + j_0(2Qr)\right)$	The EISF response predicted for a proton undergoing a jump between two equidistant sites on a circle of radius, $r(\text{Å})$.
$A_{0,\text{spherical}}(Q) = (j_0(Qr))^2$	Continuous diffusion on a spherical surface of radius, $r(\text{Å})$.
$A_{0,\text{volume}}(Q) = (3j_1(Qr)/Qr)^2$	Continuous diffusion in the interior of a spherical volume of radius, $r(\text{Å})$, as proposed by Volino and Dianoux.[17] j_1 is the first-order Bessel function: $j_1(x) = (\sin(x)/(x^2)) - (\cos(x)/(x))$.

A word regarding application of eqn (7.8) and (7.10) to the data. First, the temperature dependence of the DWF prefix needs to be ascertained. Noting that the DWF can be written,

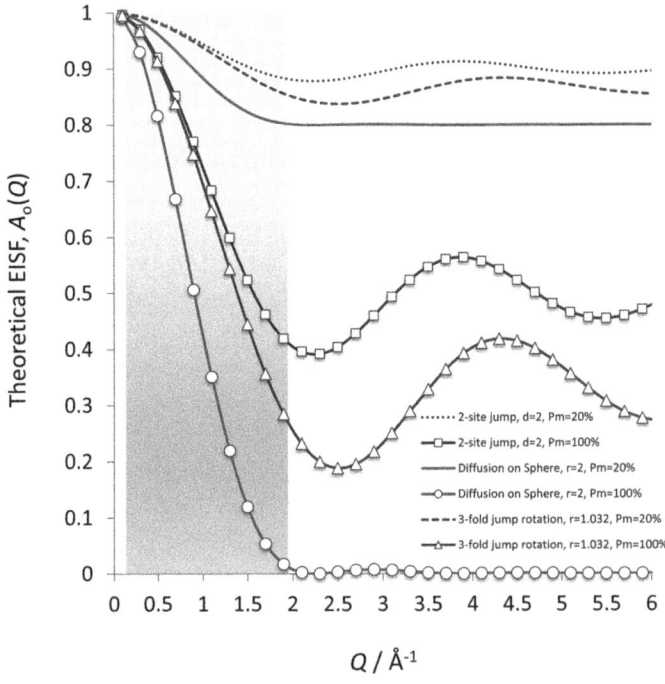

Figure 7.2 Predictive responses for three types of localised motion. The models are plotted twice: first assuming that all species are mobile and moving in a similar fashion (symbols) and, second, that just 20% of the atoms are ambulatory (lines). The shaded region depicts the typical momentum transfer window ($Q = 0.2$–1.9 Å^{-1} at the elastic line) accessible using direct/indirect QENS instrumentation.

$$\text{DWF}(Q,T) = \exp\left(-\frac{1}{3}Q^2 \frac{\mathrm{d}\langle u^2 \rangle}{\mathrm{d}T} T\right) \qquad (7.13)$$

then the rate constant, $\mathrm{d}\langle u^2 \rangle/\mathrm{d}T$, can either be considered a fit variable during the analysis of $I_i^{\text{el}}(Q,T)$ or determined experimentally and subsequently fixed. As shown in Figure 7.1(iii), $\mathrm{d}\langle u^2 \rangle/\mathrm{d}T$ is determined experimentally by fitting the rate of change of $\langle u^2(T) \rangle$ at the lowest temperatures, *i.e.* where the system can be considered a harmonic solid. $\mathrm{d}\langle u^2 \rangle/\mathrm{d}T$ is simply the gradient of a linear fit to the data.

Second, if true mobile fractions are to be extracted, line widths modelled or activation energies gleaned, then accurate knowledge of a specie's spatial constraint, *i.e.* $A_0(Q)$, is paramount. $A_0(Q)$ itself can be extracted experimentally if *a priori* information is assumed for key fit parameters. However, as one may appreciate from the forms of $A_0(Q)$ given previously, the EISF model decided upon is dependent upon the ability to observe experimentally the oscillatory nature of the Bessel function. To accurately identify radii, or jump distances, therefore, minima should fall within an instrument's accessible Q range.

Finally, while wholly mobile, yet locally constrained, molecules may allow clear distinction between different geometric behaviours, a sample exhibiting only partial mobility results in a 'muted' response, as illustrated in Figure 7.2. High statistics, detailed Q information (*i.e.* minimal detector grouping) and an instrument choice that can access a wide spatial range is therefore paramount if different $A_0(Q)$ forms be are to be compared and contrasted. In practice, using theoretical descriptions to predict possible experimental outcomes can help gauge which spectrometer to choose when planning your experiment.

As an example of using the Grapengeter model to describe $I_i^{el}(Q, T)$, normalised elastic window scan data and associated fits from lyophilized apoferritin[18] (hollow circles) at $Q = 0.744, 0.94, 1.123, 1.443$ and 1.684 Å^{-1} are shown in Figure 7.3. The data was collected using the

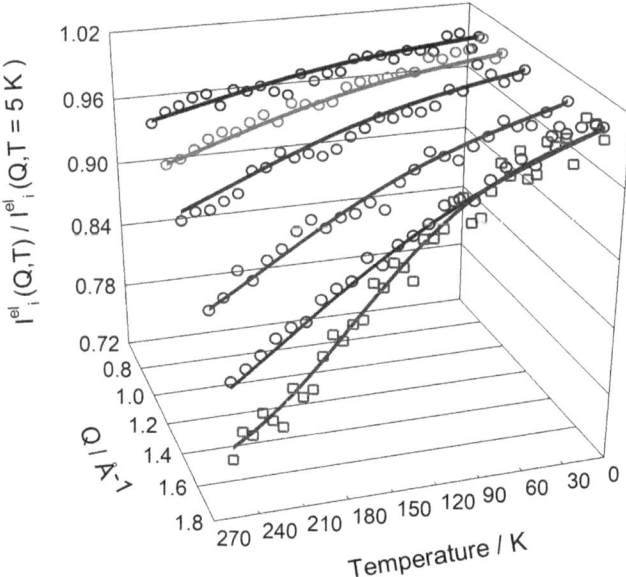

Figure 7.3 Normalised elastic incoherent window scan data from lyophilized apoferritin[18] (hollow circles) at $Q = 0.744, 0.94, 1.123, 1.443$ and 1.684 Å^{-1}. The solid lines are the result of simultaneously fitting eqn (7.12) to the data. Normalised elastic window scan data from the same sample collected using the higher energy resolution spectrometer, IN16, is shown for comparison at $Q = 1.66$ Å^{-1}. The IN16 data was also fitted using eqn (7.12).[18]

OSIRIS backscattering instrument at the ISIS Facility, UK (Γ_{res} = 24.5 µeV (FWHM)) with subsequent analysis allowing mobile fractions, p_{m}, and activation energies, E_{a}, to be extracted and compared with molecular dynamics simulations. The solid lines are the result of simultaneously fitting eqn (7.12) to the data and assuming $A_0(Q) = \frac{1}{3}\left(1 + 2j_0\left(Qr\sqrt{3}\right)\right)$. The fit parameter $d\langle u^2\rangle/dT$ was determined by modelling the rate of change of $\langle u^2(T)\rangle$ at low temperature (T < 100 K) from which a rate constant of 4.4 (± 0.65) × 10^{-4} Å^2 K^{-1} was established. Normalised elastic window scan data from the same sample collected using the higher energy resolution spectrometer, IN16 (Institut Laue–Langevin, France, Γ_{res} = 1.5 µeV (FWHM), hollow squares), and also fit using eqn (7.12), is shown for comparison at Q = 1.66 Å^{-1}. Note the effect of instrument resolution on the EFWS response; the IN16 spectrometer is designed to probe slower relaxation rates than OSIRIS.

7.2   Inelastic Fixed Window Scans (IFWS)

As mentioned in Chapter 5, a complementary method of following dynamical changes as a function of temperature is to perform an inelastic fixed window (IFWS) measurement. Unlike the EFWS approach, one chooses to follow changes in scattering intensity at a finite energy offset. As an example, for an indirect geometry spectrometer, $\mathbf{k}_{\text{f}}$ is fixed, $\mathbf{k}_{\text{i}} \neq \mathbf{k}_{\text{f}}$ and $\Delta\hbar\omega = \hbar\omega_{\text{off}} \neq 0$. Considering spatially restricted motions, and assuming again Arrhenius-type behaviour, the IFWS intensity at an energy offset, $\hbar\omega_{\text{off}}$, is expected to evolve according to,[19]

$$I_{\omega_{\text{off}}}^{\text{IFWS}}(Q,T) \propto \frac{B}{\pi}\left(1 - A_0(Q)\right)\left[\frac{\tau(T)}{1 + \omega_{\text{off}}^2\,\tau(T)^2}\right] \tag{7.14}$$

where B is a constant and τ is the relaxation time.

Analysis of IFWS profiles, when used in combination with EFWS measurements, can help unravel complex dynamical information since the IFWS probes quasi-elastic information directly and should not be contaminated by elastic scattering, background scatter or $S(\mathbf{Q})$. A typical IFWS, or $I_{\omega_{\text{off}}}^{\text{IFWS}}(Q, T)$, response can be summarised as follows.

At the lowest temperatures, relaxation rates will most likely be slower than the spectrometer's experimental observation time. Here, therefore, all spectral information is resolution limited and only elastic intensity is measured; $I_{\omega_{\text{off}}}^{\text{IFWS}}$ is zero. As T increases, however, and a faster dynamic response evolves, the quasi-elastic contribution to the scattering intensity broadens until certain energy transfer events match $\hbar\omega_{off}$. Subsequently, $I_{\omega_{off}}^{IFWS}$ starts to increase. Such an increase cannot be maintained indefinitely, however, and at the highest temperatures the QENS signal has broadened to such an extent that the intensity at $I_{\omega_{off}}^{IFWS}$ starts to decrease, before eventually plateauing and appearing as a flat background. As a result, $I_{\omega_{off}}^{IFWS}$ decreases. However, as T increases, one expects $I_{\omega_{off}}^{IFWS}$ to first pass through a maximum at,

$$T_{max} = -\frac{E_a/k}{\ln\left(\dfrac{1}{\omega_{off}\tau_0}\right)} \tag{7.15}$$

where k is the Boltzmann constant and τ_0 is the high-T relaxation limit. For systems that exhibit simple localised diffusion, the width of the quasi-elastic contribution, or correlation time, is not expected to vary with scattering vector, Q. As a result, T_{max} is seen to be Q independent. This, however, is not the case for translational motions where a change in T_{max} with momentum transfer is predicted. An example of translational diffusion in a glycerol–water mixture and as seen using the IFWS measurement method is shown in Figure 7.4(i).[19] Such distinct behaviours allow immediate, fit free identification of diffusion type. One drawback of the IFWS method, however, is that for many systems the QENS signal is *much* weaker than the elastic scattering intensity. To isolate such Q dependencies therefore, very high statistics measurements in the individual detectors are required at each temperature.

Example IFWS data sets are shown in Figure 7.4 from the study of (i) glycerol in water, (ii) the connection between fragility, mean squared displacement and shear modulus in the van der Waals bonded glass-forming liquid, cumene,[21] and (iii) hydrogen dynamics in the low temperature phase of lithium borohydride (LiBH$_4$).[20] Considering the LiBH$_4$ example, you will notice that while peak intensity is at a maximum at 210 K, the onset of this peak is closer to 120 K and correlates with a decrease in elastic scattering intensity. When processing IFWS data, therefore, it is important to note whether or not the energy offset used for the measurement falls within the resolution of the neutron instrument. If so, one will expect an elastic scattering contribution to the raw IFWS data. As illustrated in Figure 7.4(iii), the $\hbar\omega_{off} = 2$ µeV energy offset used for the LiBH$_4$ study sits in the wings of the IN16 instrument's resolution function (inset). Visual inspection shows the raw IFWS data to exhibit an irregular temperature dependent background response that mimics the temperature dependence of the EFWS signal. In practice, therefore, IFWS data should be corrected for elastic scattering contamination. Such contamination can

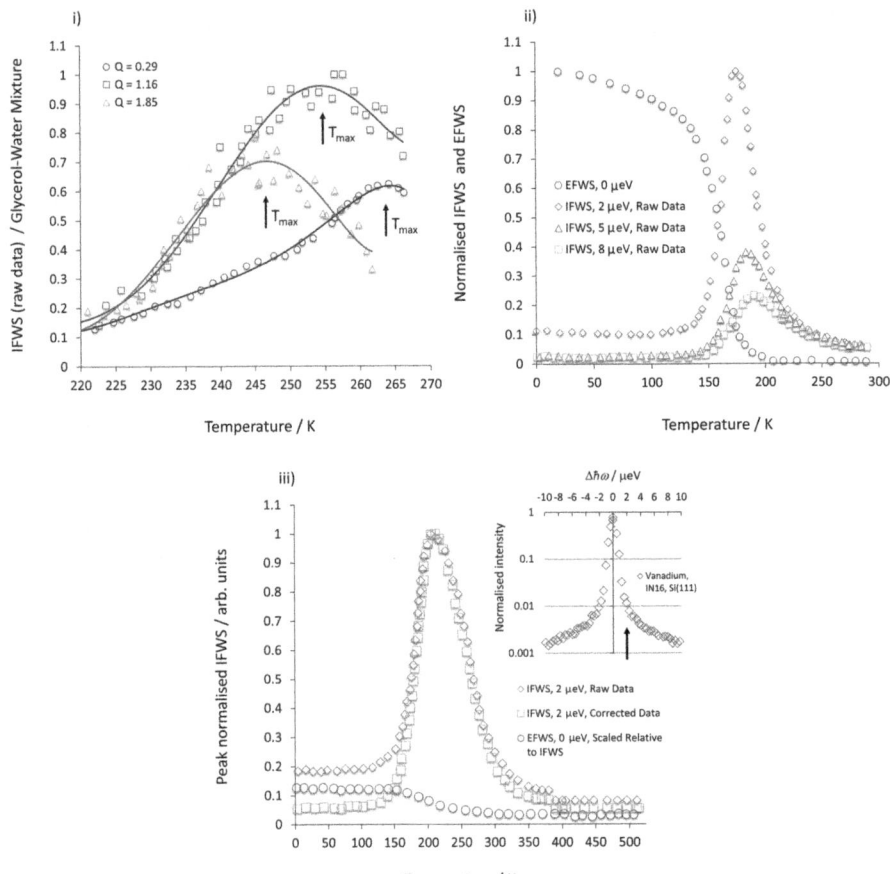

Figure 7.4   (i) A study of glycerol in water. Data points extracted from Frick
et al.[19] Note the Q dependence of the IFWS peak, a response
characteristic of translational diffusion. The lines are a guide to
the eye. (ii) An investigation to connect fragility, mean squared
displacement and shear modulus in the van der Waals bonded
glass-forming liquid, cumene. Data extracted from Hansen *et
al.*[21] The IFWS and EFWS data sets were collected concurrently
using the IN16B instrument configured to measure at $\hbar\omega_{off}$ = 2,
5 and 8 μeV. Since the raw data shown is the result of summing
neutron counts across the detector bank, no Q dependencies
can be inferred. α-relaxation is seen entering the instrument
window around 150 K. Note the decrease in elastic contribution
to the IFWS signal at low temperatures with increasing $\hbar\omega_{off}$. (iii)
The study of hydrogen dynamics in the low temperature phase
of lithium borohydride ($LiBH_4$). Data extracted from Remhof *et
al.*[20] The IFWS and EFWS data sets were collected using the
IN16 instrument configured to measure at $\hbar\omega_{off}$ = 2 μeV. Since
the raw data shown is the result of summing neutron counts
across the detector bank, no Q dependencies can be inferred.
The drop in intensity in both the IFWS and EFWS at 380 K is
associated with a solid–solid structural phase. Inset: the IN16
resolution function showing that an offset energy of $\hbar\omega_{off}$ = 2
μeV sits within its wings.

be removed by collecting EFWS and IFWS data concurrently at each temperature, scaling the elastic signal relative to the IFWS intensity and subtracting. The benefit of such subtraction is clearly visible in Figure 7.4(iii). It should also be noted that contamination from elastic scattering becomes less of an issue the further away from the elastic line you measure, as illustrated in Figure 7.4(ii).

References

1. V. Arrighi, D. Batt-Coutrot, C. Zhang, M. T. F. Telling and A. Triolo, *J. Chem. Phys.*, 2003, **119**, 1271–1278.
2. W. Doster, *Eur. Biophys. J. Biophys. Lett.*, 2008, **37**, 591–602.
3. M. T. F. Telling, C. Neylon, S. H. Kilcoyne and V. Arrighi, *J. Phys. Chem. B*, 2008, **112**, 10873–10878.
4. W. Doster, H. Nakagawa and M. S. Appavou, *J. Chem. Phys.*, 2013, **139**, 045105.
5. S. Gabrielli, *Università degli Studi di Roma Tor Vergata*, 2019.
6. A. Rahman, K. S. Singwi and A. Sjolander, *Phys. Rev.*, 1962, **126**, 997–1004.
7. D. Zeller, M. T. F. Telling, M. Zamponi, V. G. Sakai and J. Peters, *J. Chem. Phys.*, 2018, **149**, 234908.
8. J. Peters and G. R. Kneller, *J. Chem. Phys.*, 2013, **139**, 165102.
9. Z. Yi, Y. L. Miao, J. Baudry, N. Jain and J. C. Smith, *J. Phys. Chem. B*, 2012, **116**, 5028–5036.
10. T. Becker and J. C. Smith, *Phys. Rev. E: Stat., Nonlinear, Soft Matter Phys.*, 2003, **67**, 021904.
11. W. Doster, S. Cusack and W. Petry, *Nature*, 1989, **337**, 754–756.
12. G. Zaccai, *Science*, 2000, **288**, 1604–1607.
13. G. Zaccai, F. Natali, J. Peters, M. Řihová, E. Zimmerman, J. Ollivier, J. Combet, M.-C. Maurel, A. Bashan and A. Yonath, *Sci. Rep.*, 2016, **6**, 37138.
14. H. H. Grapengeter, B. Alefeld and R. Kosfeld, *Colloid Polym. Sci.*, 1987, **265**, 226–233.
15. V. Arrighi, J. S. Higgins, A. N. Burgess and W. S. Howells, *Macromolecules*, 1995, **28**, 4622–4630.
16. M. Bée, *Quasi-elastic Neutron Scattering: Principles and Applications in Soild State Chemistry, Biology and Materials Science*, Adam Hilger, Bristol, England, 1988.
17. F. Volino and A. J. Dianoux, *Mol. Phys.*, 1980, **41**, 271–279.
18. M. T. F. Telling, C. Neylon, L. Clifton, S. Howells, L. van Eijck and V. Garcia Sakai, *Soft Matter*, 2011, 7, 6934–6941.
19. B. Frick, J. Combet and L. van Eijck, *Nucl. Instrum. Methods Phys. Res., Sect. A*, 2012, **669**, 7–13.
20. A. Remhof, A. Züttel, T. Ramirez-Cuesta, V. Garcia-Sakai and B. Frick, *Chem. Phys.*, 2013, **427**, 18–21.
21. H. W. Hansen, B. Frick, T. Hecksher, J. C. Dyre and K. Niss, *Phys. Rev. B*, 2017, **95**, 104202.

8 $S(\boldsymbol{Q},\omega)$ and $I(\boldsymbol{Q},t)$

In this chapter we will consider:

- The analysis of the dynamic structure factor, $S(\boldsymbol{Q},\omega)$.
- The analysis of the intermediate scattering function, $I(\boldsymbol{Q},t)$.

8.1 Analysis of the Dynamic Structure Factor, $S(\boldsymbol{Q},\omega)$

At its simplest, analysis of EFWS and/or IFWS data informs, *via* inflexions or peaks, where quasi-elastic processes enter, and leave, the observable time window of a neutron instrument. It thus follows to explore these regions in greater detail to extract further descriptive parameters *via* analysis of the dynamic structure factor, $S(\boldsymbol{Q},\omega)$.

As discussed in Part 1, $S(\boldsymbol{Q},\omega)$ comprises both coherent and incoherent contributions weighted by their respective cross sections,

$$S(\boldsymbol{Q}, \omega) = [\sigma_c S_c(\boldsymbol{Q}, \omega) + \sigma_i S_i(\boldsymbol{Q}, \omega)] \tag{8.1}$$

However, in truth, since a large number of $S(\boldsymbol{Q},\omega)$ experiments are interested in the exploration of self (*i.e.* $S_i(\boldsymbol{Q},\omega)$), rather than collective (*i.e.* $S_c(\boldsymbol{Q},\omega)$), motion(s), samples are prepared such that the measured signal is dominated by incoherent, rather than coherent, scattering.

As way of illustration, evolution of $S_i(\boldsymbol{Q},\omega)$ as a function of scattering angle, hence Q, is presented in Figure 2.1(i). By modelling such variation in terms of line width and peak intensity, information about the

A Practical Guide to Quasi-elastic Neutron Scattering
By Mark T. F. Telling
© Mark T. F. Telling 2020
Published by the Royal Society of Chemistry, www.rsc.org

frequency, and possibly geometry, of the excitation(s) probed can be extracted. Clearly, there are several approaches to line width analysis, the methodology chosen being a matter of personal preference that can include basic least squares fitting or more extensive Bayesian analysis.[1,2] However, when working in energy space, and regardless of the technique eventually adopted, you will most likely start your analysis by modelling the measured dynamic structure factor in terms of Lorentzian functions convolved with the resolution function of the neutron instrument used,

$$S_i^{\text{expt}}(\boldsymbol{Q},\omega) = R(\boldsymbol{Q},\omega) \otimes S_i(\boldsymbol{Q},\omega) \tag{8.2}$$

an approach that allows the number of dynamic processes contributing to $S_i(\boldsymbol{Q},\omega)$ to be identified alongside the Q and temperature dependencies of the respective Lorentzian function line widths and intensities. If we recall that $S_i(\boldsymbol{Q},\omega)$ can be written in terms of simply quasi-elastic *or* elastic and quasi-elastic incoherent structure factors, *i.e.*,

$$S_i(\boldsymbol{Q},\omega) = \exp\left(-\frac{Q^2 <u^2(T)>}{3}\right)\left(A_0(\boldsymbol{Q})\delta(\omega) + A_1 S_i^{\text{qe}}(\boldsymbol{Q},\omega)\right) \tag{8.3}$$

then the incoherent dynamic structure factor may comprise one or a combination of the following spectral components:

- Wholly elastic scattering.
  - *i.e.* the motion(s) of all the atoms probed are slower than the experimental time window of the spectrometer used. The motion(s) are said to be resolution limited.

- A broad 'flat background'-like response.
  - *i.e.* the motion(s) of all the atoms probed are faster than the experimental time window of the spectrometer used.

- Quasi-elastic broadening(s) only.
  - *i.e.* the motion(s) of all atoms are within the spectrometer's observable time window and the system exhibits translational diffusion.

- Quasi-elastic *and* elastic scattering intensity.
  - *i.e.* the motion(s) of all atoms are within the spectrometer's observable time window but are geometrically constrained (localised diffusion).

- A combination of the above.
  - *i.e.* the relative intensity of the wholly elastic signal being used to determine an immobile fraction.

It is worth commenting that rarely can more than two distinct dynamic processes be isolated with accuracy from an experimentally determined $S_i(Q,\omega)$ data set using simple least squares fitting. If more than two processes are expected then Bayesian probability analysis may prove a more reliable analysis tool. Alternatively, interpretation may require a justified theoretical model and/or supporting information from complementary methods or other neutron spectrometers that straddle different dynamic ranges. Additionally, and as a rule of thumb, distinct dynamic contributions to $S_i(Q,\omega)$ will show marked difference in line width (usually a decade). Analysis that suggests the presence of two processes with comparable line widths should be questioned.

Monitoring the Q dependence of a quasi-elastic scattering line width allows diffusion coefficients, residence times and jump lengths to be ascertained. Moreover, modelling a line width as a function of temperature, at a specific Q, may enable Arrhenius behaviour (Figure 8.5) to be observed and activation energies to be determined. Indeed, line width analysis reveals two general types of motion; translational (long range, unhindered) or localised (geometrically constrained). As such, the total incoherent dynamic structure factor may well consist of signals from both,

$$S_i(Q,\omega) = \exp\left(-\frac{Q^2 <u^2(T)>}{3}\right)\left(S_i^R(Q,\omega) \otimes S_i^T(Q,\omega)\right) \qquad (8.4)$$

An example of modelling $S_i(Q,\omega)$ collected using the FOCUS instrument (Paul Scherrer Institut, Switzerland, 290 K, $Q = 0.55$ Å^{-1}) is shown in Figure 8.1. Here, least squares analysis of the measured incoherent dynamic structure factor allowed the spectral components associated with translational, $S_i^T(Q,\omega)$, and rotational, $S_i^R(Q,\omega)$, bulk cytoplasmic H_2O dynamics in red blood cells (RBC) to be isolated.[3] This work also aimed to explore interfacial water mobility. However, being slower than the observable time window on FOCUS, the signal from the haemoglobin hydration shell presented itself as an elastic scattering contribution. The Q dependence of the Lorentzian line width associated with the translational diffusion process, and as followed using the IRIS instrument, is shown in the inset; additional measurements using the higher energy resolution of IRIS being performed in an attempt to resolve the elastic scattering signal observed using FOCUS.

The following discussion will decouple $S_i^T(Q,\omega)$ and $S_i^R(Q,\omega)$ and consider what one might expect to observe experimentally for each. It should be noted that the following discussion does not include all possible diffusion models. For a more expansive overview, and for

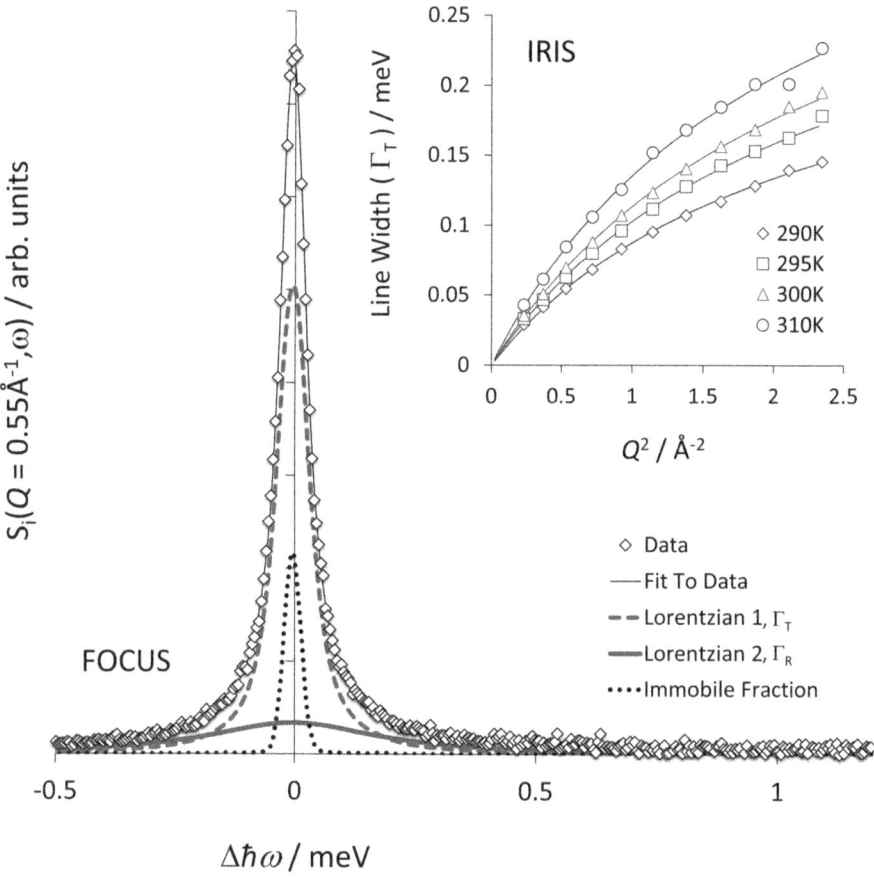

Figure 8.1 Least squares analysis of $S_i(\mathbf{Q},\omega)$ recorded using the FOCUS instrument (PSI, 290 K, $Q = 0.55$ Å^{-1}) and showing the spectral components associated with translational and rotational bulk cytoplasmic H_2O dynamics. An immobile (elastic scattering) fraction is also observed that is attributed to the haemoglobin hydration shell. Data extracted from Stadler *et al.*[3] Inset: the Q dependence of the Lorentzian line width (Γ_T, HWHM) associated with translational diffusion. The solid lines are fits using the jump-diffusion line width model of Singwi and Sjölander[4] from which a translational diffusion coefficient, D, of 2×10^{-5} cm^2 s^{-1} was obtained.

models that are perhaps better aligned to your particular system, you are referred to Bée[5] and/or the scientific literature.

8.1.1 Translational Diffusion

Let's first consider continuous long-range isotropic translational diffusion alone. In this case, particles are considered to be unconstrained, especially compared to the spatial range probed by the neutron instrument, and move linearly under the influence of the collisions between them. Collision results in a random change of direction with no memory of prior events. Here, particle concentration fluctuations are described by Fick's law,[6] with the incoherent dynamic structure factor and associated self-intermediate scattering function being described using simply a Lorentzian or exponential, respectively. In this case, the Q dependence of the quasi-elastic broadening (at a given temperature) can be related to the particle's diffusion coefficient *via*,[7]

$$\Gamma(\text{HWHM}) = DQ^2 \qquad (8.5)$$

where Γ is the line broadening at half width at half maximum (HWHM) and D (m^2 s^{-1}) is the diffusion coefficient. A signature of continuous diffusion, therefore, is a linear response if Γ is plotted as a function of Q^2 with the gradient being equal to D.

Of course, most systems studied today are not so 'simple' but instead exhibit dynamical responses that result in line width behaviour that deviates from DQ^2. As an example, continuous diffusion may be hindered by hydrogen-bonding phenomena that impart a period of residence between jumps. Revised line width models (aka jump diffusion models) are therefore proposed in the literature that consider deviations from the continuous diffusion response at small length scales, *i.e.* large Q. Some of the more common predictive forms are listed in Table 8.1 and illustrated in Figure 8.2. However, all exhibit DQ^2 behaviour at low Q and then approach an asymptote related to $1/\tau_{\text{jump}}$ at elevated momentum transfer. An example of jump diffusion behaviour as seen from water confined in the pores of the chemically cross-linked hydrogel, poly(vinyl alcohol), is shown in Figure 8.2 (inset).[8] The jump-diffusion model of Singwi and Sjölander,[4] as used to describe the Q dependence of Γ_{T} associated with the translational diffusion of bulk cytoplasmic H$_2$O in RBC, is shown in Figure 8.1.

8.1.2 Localised Diffusion

What happens, however, if our diffusing entity is no longer free to move without constraint, but is geometrically anchored. Such motion, which can be either continuous or defined in terms of jumps between specific sites, is referred to as localised. In this case we observe both an elastic scattering signal plus a quasi-elastic response. Analysis is performed in the same manner as discussed above but, unlike translational diffusion, a *Q independent line width* is observed. Modelling the associated elastic incoherent structure factor (EISF) yields not only information about the geometry of said motion but also the

Table 8.1 Predictive line width models used to describe continuous and jump diffusion phenomena.

Model	Function ($\Gamma(\mathbf{Q})$, HWHM)
1. *Continuous diffusion*[6] Developed to predict $\Gamma(Q)$ for particles undergoing long range continuous (Brownian) diffusion in a macroscopic medium that is described by Fick's law. $\langle l^2 \rangle$ = *the* mean square jump length/displacement. τ = time between collisions. D (m^2 s^{-1}) = the diffusion coefficient.	$= DQ^2 = \dfrac{\langle l^2 \rangle}{6\tau} Q^2$
2. *Chudley–Elliot jump diffusion (1961)*[9] Developed considering a liquid in its quasi-crystalline form. Has found application for the study of atoms diffusing through a lattice. l = jump length between sites, τ_0 = mean residence time.	$= \dfrac{1}{\tau_o}\left(1 - \dfrac{\sin(Ql)}{Ql}\right)$
3. *Hall and Ross jump diffusion (1981)*[10] Developed for random jump diffusion with a Gaussian distribution of jump lengths. τ = mean jump rate. $\langle l^2 \rangle$ = the mean jump length.	$= \dfrac{1}{\tau}\left[1 - \exp\left(-\dfrac{Q^2 \langle l^2 \rangle}{6}\right)\right]$
4. *Singwi and Sjölander jump diffusion (1960)*[4] Developed for liquid water. The diffusion model proposed hypothesises that each molecule oscillates for a mean time, τ_0, before undergoing continuous linear diffusive motion. $\langle R^2 \rangle$ = the mean square radius of the thermal cloud developed during oscillatory motion. $\langle l^2 \rangle$ = the mean jump length associated with the continuous diffusion.	$= \dfrac{1}{\tau_0}\left[1 - \left(\dfrac{\exp(-2W)}{1 + DQ^2\tau_0}\right)\right]$ where $2W = DQ^2\tau_0 \dfrac{\langle R^2 \rangle}{\langle l^2 \rangle}$
5. *Teixeira water jump diffusion (1985)*[7] Developed following observation of both rotational and translational diffusion on different time scales. Complete decoupling of the motion types is assumed with the measured scattering function being expressed as the convolution of $S_i^R(\mathbf{Q}, \omega)$ and $S_i^T(\mathbf{Q}, \omega)$. Super-cooling the water allowed rotational dynamics to be modelled alone. See also Section 8.4.	$= \dfrac{DQ^2}{1 + DQ^2\tau_0};\quad D = \dfrac{\langle l^2 \rangle_{av}}{6\tau_0}$

percentage of mobile species contributing to the quasi-elastic scattering signal. To recap, the EISF can be calculated from the relative intensities, and Q variation, of the elastic and quasi-elastic scattering signals *via*,

$$\text{EISF} = A_0(Q) = \frac{I^{el}(Q)}{I^{el}(Q) + I^{qe}(Q)} \tag{8.6}$$

For a system exhibiting a single localised process, the structure factor, A_1, associated with the intensity of the quasi-elastic scattering contribution follows a Q-dependence that can be written, $A_1 = 1 - A_0$, since, at each Q, $\int_{-\infty}^{\infty} S_i(\mathbf{Q},\omega)d\omega = 1$. The form of the EISF is therefore

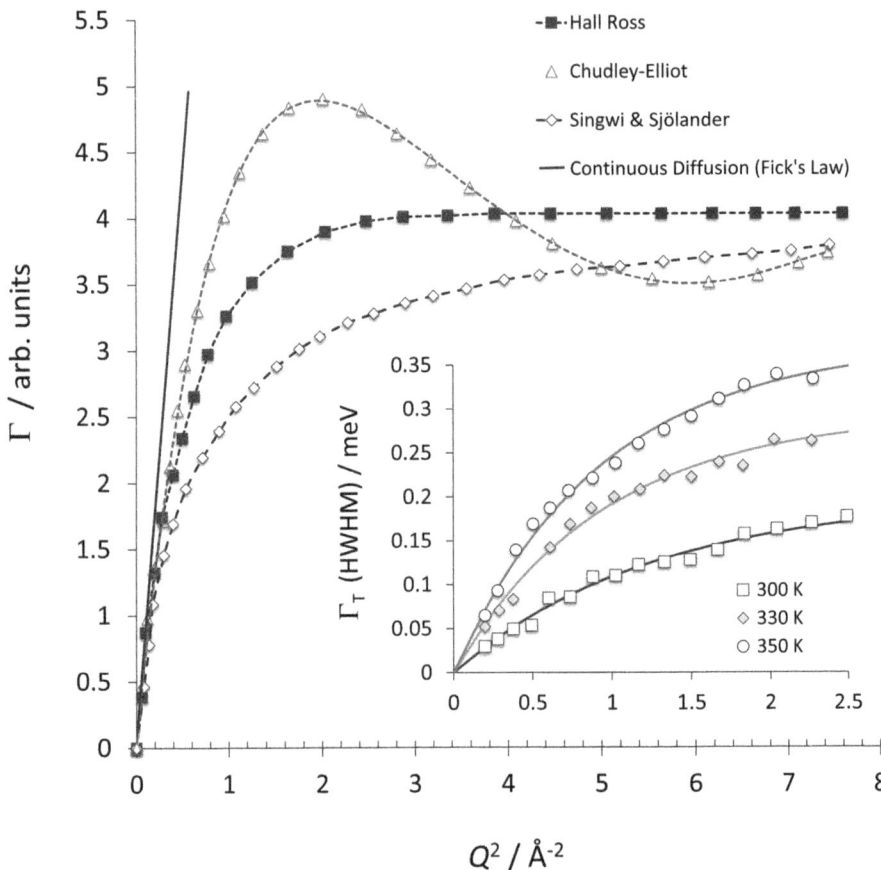

Figure 8.2 Predictive line width models used to describe: continuous diffusion, Chudley–Elliot jump diffusion, Hall–Ross Gaussian jump diffusion, Singwi and Sjölander water jump diffusion and Teixeira water jump diffusion. One notices that all jump related models predict that Γ exhibits DQ^2 behaviour at low Q and then approach $1/\tau_{jump}$ at elevated Q. Inset: jump diffusion as seen from water confined in the pores of the chemically cross-linked hydrogel, poly(vinyl alcohol). Line widths extracted from Paradossi *et al.*[8] and described (lines) using the unrestricted jump model of Singwi and Sjölander.[4]

given by the energy integrated area of the elastic scattering response divided by the energy integrated area of the total scattering, *i.e.* eqn (8.6), with its Q dependence denoting a specific geometry of motion. As we have discussed, the motion described is revealed by comparing experimentally determined EISF values to theoretical predictions.

8.1.3 Jump Restricted Diffusion

When dealing with a system in which jump diffusion occurs within a spatially restricted environment then we may encounter a situation where both Q-dependent and Q-independent line width behaviours co-exists. If, for example, we assume jump diffusion in one dimension between two walls separated by distance, L, then the Q-dependence of the quasi-elastic scattering line width for random jump restricted diffusion will exhibit asymptotic behaviours at both low and high Q, *i.e.* characteristics of diffusion within a restricted volume at low-Q and that of a jump diffusion model at high-Q. We assume here, of course, that L will at some point become smaller than the spatial range probed by the neutron instrument. Such limits can be explained as follows. At low Q we are concerned with large length scales and thus the strong influence of narrow boundaries on the diffusion process. Here the line width deviates from the DQ^2 law and tends to a finite Γ value with the effective linear dimension of the confining medium being inferred from the upper Q limit of the line width plateau; $Q_{upper} \sim 2\pi/L_{conf}$ (Figure 8.2). In contrast, at large momentum transfer values, motion over short distances dominates ($l \ll L_{conf}$) and, because the elementary displacement of the particles in not infinitely small, the HWHM of the quasi-elastic component tends to $1/\tau_{jump}$. Between these two limits the DQ^2 response is recovered since for small distances, in comparison to the confining medium, the dimension of the boundaries becomes insignificant. An example of random jump restricted diffusion as seen for ammonia in NH_3-SCR zeolite catalysts is shown in Figure 8.3.

8.2 Susceptibility, $\chi''(\mathbf{Q},v)$

As mentioned, modelling $S_i(\mathbf{Q},\omega)$ is often approximated *via* the convolution of Lorentzian functions, *i.e.* the supposition of a finite number of distinct exponential decays. In dynamically complex systems, however, in which a distribution of relaxation processes exists, such approximation is often misleading. Instead, heterogeneous environments may be better described in terms of either dynamical susceptibility in the frequency domain (*i.e.* $\chi''(v) = (1 - iv\tau)^{-b}$ with $b < 1$) and described using a Cole–Davidson distribution function, as a stretched exponential in the time domain described by a Kohlrausch–Williams–Watts equation (*i.e.* $\exp(-(t/\tau)^\beta)$ with $\beta \leq 1$) or *via* application of the Havriliak–Negami relaxation model.

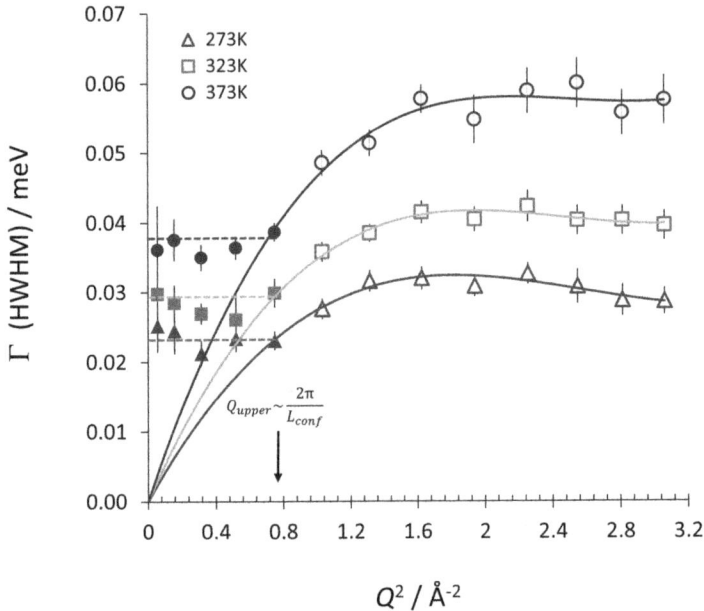

Figure 8.3   Random jump restricted diffusion as seen for ammonia in NH_3-SCR zeolite catalysts. Two distinct Q regimes are observed. At high Q (>1 Å^{-1}) the data (open symbols) can be fit using the Chudley–Elliot jump diffusion model. However, at low Q the line width follows the general formalism for diffusion inside a confined volume (filled symbols). Data extracted from O'Malley.[11] The HWHM plateau cut-off is at approximate Q_{upper} = 0.87 Å^{-1}, giving an effective confinement length of L_{conf} = 7.2 Å^{-1}.

Considering the former, if we convert the incoherent dynamic structure factor to frequency space, *i.e.* $S_i(Q,v)$, then the dynamic susceptibility, $\chi''(Q,v)$, is related to $S_i(Q,v)$ through the Bose occupation number $n_B(v) = (\exp(hv/kT) - 1)^{-1}$ such that,

$$\chi''(Q, v) \propto S_i(Q, v)/n_B(v) \qquad (8.7)$$

The advantages of using susceptibility analysis to probe complex relaxation phenomena are discussed by Roh *et al.*[12] and include:

- Scattering data can be directly compared to dielectric, $\varepsilon''(v)$ or mechanical loss data, $G''(v)$.
- Relaxation processes appear as a maximum at $v_{max} \sim 1/(2\pi\tau)$. It is therefore easy to distinguish between slow and fast processes since relaxation processes with well-separated relaxation rates, τ, appear as separated peaks.

- The spectral shape provides information about the distributed nature of the relaxation phenomenon, *i.e.* values of $b = 1$ (or −1) pointing to a lone relaxation rate.
- Temperature variation is accounted for.

The wave vector dependence of $\chi''(v)$ is determined by fitting the distribution function,[13]

$$\chi''(v) = \chi_0 \frac{\left(\dfrac{v}{v_{max}}\right)^{1-\alpha} \cos\left(\dfrac{\pi\alpha}{2}\right)}{1+2\left(\dfrac{v}{v_{max}}\right)^{1-\alpha} \sin\left(\dfrac{\pi\alpha}{2}\right)+\left(\dfrac{v}{v_{max}}\right)^{2-2\alpha}} \tag{8.8}$$

v_{max}, α and χ_0 are fit parameters. The above form is traditionally used to describe symmetrically stretched relaxation peaks. Susceptibility data, as generated from dynamic incoherent structure factor measurements, is shown in Figure 8.4.[14] The collated data sets were collected

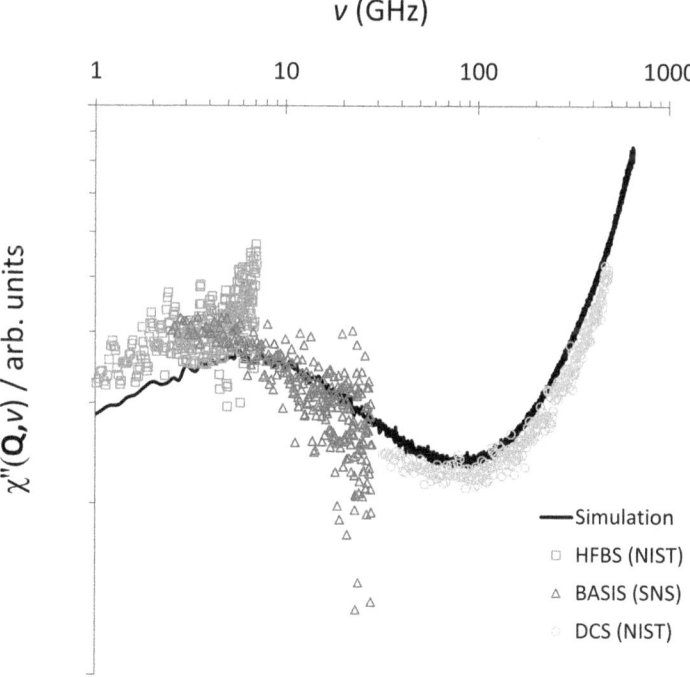

Figure 8.4 Susceptibility spectra, $\chi''(\mathbf{Q},v)$, as generated from $S_i(\mathbf{Q},v)$ neutron scattering data collected from hydrated lysozyme. Data extracted from Hong *et al.*[14]

from hydrated lysozyme ($Q_{ave} = 1$ Å^{-1}) using the instruments HFBS (indirect, NIST, USA), BASIS (indirect, SNS, USA) and DCS (direct, NIST, USA). The experimental response is compared with that predicted from a molecular dynamics simulation.

8.3 Analysis of the Intermediate Scattering Function, *I*(**Q**,*t*)

An alternative method of extracting spectral information from $S_i(Q, \omega)$ involves Fourier transforming the measured dynamic incoherent structure factor and analysing the associated intermediate scattering function, $I_s(Q,t)$, or more commonly the normalised intermediate scattering function, $I_s(Q,t)/I_s(Q,t = 0)$, in the time domain. Such an approach is advantageous for several reasons. First, Fourier transforming $S_i^{expt}(Q,t)$ removes any instrument resolution contribution. As a result, least squares model fitting of $I_s(Q,t)$ is generally straightforward. Second, and as mentioned previously, dynamically complex systems or heterogeneous environments exhibit a distribution of proton mobilities, relaxation times, jump lengths and activation energies. We cannot, therefore, adequately describe such complexity using just one or two Lorentzian functions. If dynamical susceptibility analysis is not an option, then analysis in the time domain often proves advantageous. Third, for spectrometers operating in Q–ω space, the measured structure factor is a convolution of spectral contributions (eqn (2.23)). In contrast, in the time domain the measured response is simply a product of all dynamical components (eqn (2.24)). Finally, analysis in the time domain allows data to be collated across all quasielastic instrument classes. $S_i(Q,\omega)$ measured on a in/direct geometry instrument and transformed can be likened to $I_s(Q,t)$ recorded directly on an NSE instrument, *assuming the two data sets have been collected at the same momentum transfer.* As such, relaxation phenomena can be tracked over a wider time window.

For a single relaxation process, the normalised self-intermediate scattering function will manifest itself as a simple exponential,

$$\frac{I(Q,t)}{I(Q,t=0)} = \text{DWF} \times \exp\left(-\left(\frac{t}{\tau}\right)\right) \tag{8.9}$$

Here, τ is the relaxation time. A system that exhibits a distribution of relaxation rates, however, may, as mentioned, be better described using the Kohlrausch–Williams–Watt (KWW), aka stretched exponential form, *i.e.*,

$$\frac{I(Q,t)}{I(Q,t=0)} = \text{DWF} \times \exp\left(-\left(\frac{t}{\tau_{\text{KWW}}}\right)^{\beta}\right) \tag{8.10}$$

Non-exponential behaviour in eqn (8.10) is immediately apparent if the stretching parameter falls below unity; β assuming a value between zero and one. Should the motion probed be both unique and localised then eqn (8.9) is modified such that it includes the elastic incoherent structure factor, $A_0(Q)$,

$$\frac{I(Q,t)}{I(Q,t=0)} = \text{DWF} \times \left[A_0(Q) + \left[\left[1 - A_0(Q)\right]\exp\left(-\left(\frac{t}{\tau}\right)\right)\right]\right] \tag{8.11}$$

The EISF, $A_0(Q)$, itself can be extracted by considering the Q dependence of the plateau intensity reached in the long time limit, *i.e.* as $I_s(Q,t) \rightarrow \infty$. Of course, for a system showing partial mobility, p_m, an effective EISF, $A_0'(Q)$, will be extracted. Similar modification can be made to eqn (8.10). It should also be noted that τ_{KWW} is an effective relaxation time that is dependent upon both T and β, or more correctly the temperature dependence of the spectral shape of the distribution. As discussed by, for example, Arbe *et al.*[15] and Tanchawanich *et al.*,[16] a mean relaxation time, $\langle\tau\rangle$, can be extracted using the relationship,

$$\langle\tau\rangle = \Gamma_G\left(\frac{1}{\beta}\right)\frac{\tau_{\text{KWW}}}{\beta} \tag{8.12}$$

which gives an alternative, more representative (*i.e.* decoupled) temporal response from which a mean quasi-elastic scattering line width can be ascertained, $\langle\Gamma\rangle$. Note that in the above equation $\Gamma_G()$ is the *gamma function* and τ is recovered for $\beta = 1$.

For systems exhibiting a distribution of relaxation rates, the merits of transforming and analysing $S_i(Q,\omega)$ data in the time domain are discussed by Howells.[17] As shown in Figure 8.5, such an approach proves most advantageous when trying to match data sets collected at a common momentum transfer vector, Q, but from neutron instruments with different energy resolutions.

For $I(Q,t)$ data measured directly using NSE, the same analysis considerations and models described above may be applied. As way of illustration, normalised NSE spectra from a concentrated protein solution measured at room temperature is shown in Figure 8.6(i). The data was modelled using the single exponential relaxation function (eqn (8.9)) described above with the inset illustrating the resulting relaxation times as a function of concentration. The lines are linear

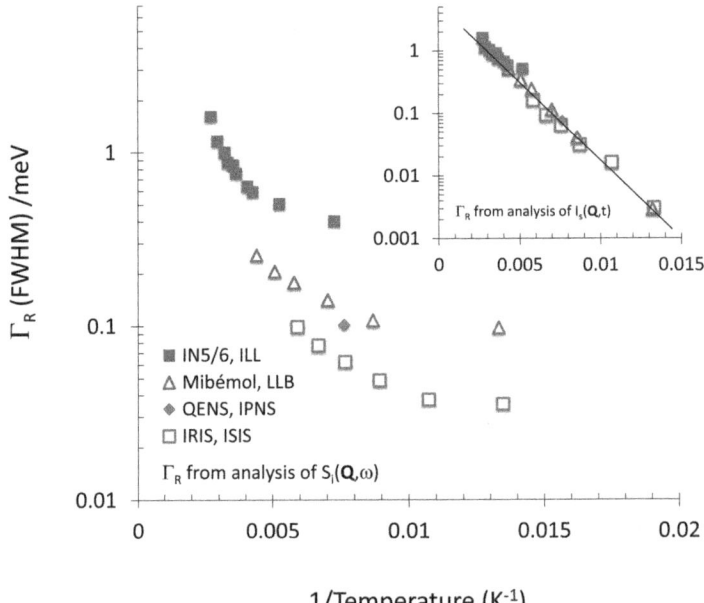

Figure 8.5 The result of $S_i(\mathbf{Q},\omega)$ line width analysis (FWHM *versus* inverse temperature) from polymethyl methacrylate (PMMA) as measured on IN5 (ILL), Mibémol (LLB), QENS (IPNS) and IRIS (ISIS). All data sets were modelled using a single Lorentzian dynamic structure factor. Inset: analysis of the associated intermediate scattering functions as modelled using the Kohlrausch–Williams–Watt form. Analysis in the time domain reveals consistent behaviour between spectrometers with the temperature dependence of polymer relaxation being seen to be Arrhenius. Data extracted from Howells.[17]

fits to the data and give global diffusion coefficients of $D_{dil} = 1.03 \times 10^{-6}$ cm^2 s^{-1} and $D_{conc} = 0.67 \times 10^{-6}$ cm^2 s^{-1}. Of course, NSE measurements are in truth more focused toward collective dynamics, $I_c(\mathbf{Q},t)$, rather than self diffusion, $I_s(\mathbf{Q}, t)$, with NSE being used extensively to study co-operative behaviour associated with, for example, polymer reptation[18] and Rouse dynamics[19] as well as for extracting information regarding membrane's stiffness.[20]

Clearly, eqn (8.9) and (8.10) are not the only functional forms used to model neutron spin echo data. While an expansive number of examples can be found in the scientific literature, as way of illustration $I(\mathbf{Q},t)$ might also be modelled using the Weron[21] or Zilman and Granek[22] equations.

Briefly, the Weron approach considers a hierarchical progression of relaxation introduced through the formation of finite clusters and was developed to explain the apparently universal power law for dielectric relaxation. The model accounts for the effects of both inter-cluster and intra-cluster interactions with the characteristic time scale of any relaxing entity being restricted by the structural reorganisation of the surrounding clusters. The resulting generalised relaxation equation takes the form,

$$\varphi(t) = \left[1 + k\left(\frac{t}{\tau}\right)^{\beta} \right]^{-1/k} \tag{8.13}$$

where β is associated with the fractal geometry of the system, τ is the relaxation time and k (>0) is an effective interaction parameter. The Weron power law reduces to the Kohlrausch form in the limit $k \to 0$. Here, $0 \leq \beta \leq 1$ has the same meaning as previously mentioned with the limit $\beta = 1$ implying simple Debye (exponential) relaxation. While developed to explain dialectic data, the Weron function has been used successfully to model the evolution of the normalised intermediate scattering function recorded from cluster glass-like materials (Figure 8.6(ii)). The spectra shown were collected using the IN11C spectrometer (Institut Laue-Langevin, France) at $Q = 0.4$ Å^{-1} with an

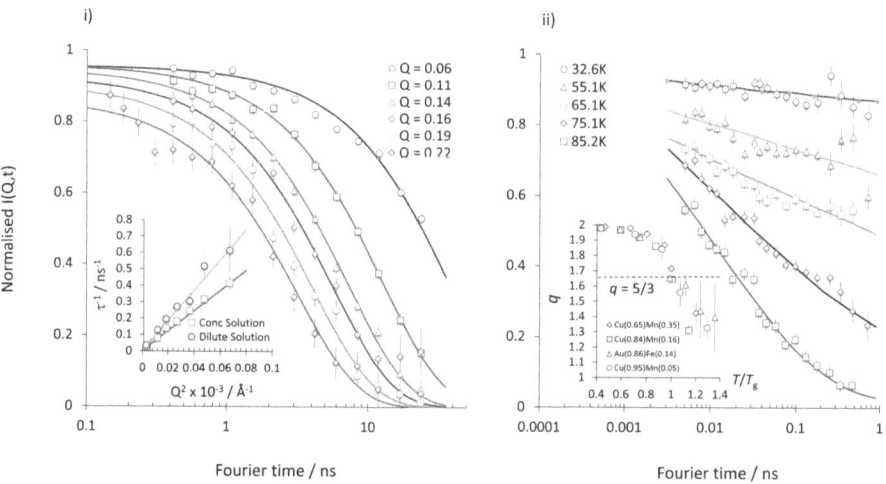

Figure 8.6 Illustrative neutron spin echo data sets and associated fits. (i) Normalised $I(Q,t)$ spectra from a concentrated protein solution measured at room temperature. The data was fitted using a single exponential relaxation function. Data points extracted from Wood *et al.*[24] (ii) The temperature dependence of NSE spectra collected from the spin glass-like material, Cu$_{1-x}$Mn$_x$, and modelled using the Weron function. Data points extracted from Pickup *et al.*[25]

incoming wavelength of 5.5 Å. The inset shows the extracted Tsallis sub-extensivity parameter, q, as a function of reduced temperature.

In contrast, the Zilman and Granek model was developed to describe membrane thermal undulations by considering an ensemble of membrane plaquettes at random orientations. The Zilman and Granek model predicts that,

$$\frac{I(Q,t)}{I(Q,t=0)} \sim \exp^{-(\Gamma_q t)^{\frac{2}{3}}} \qquad (8.14)$$

where,

$$\Gamma_q \propto \kappa^{-\frac{1}{2}} Q^3 \qquad (8.15)$$

and K is the membrane bending modulus.

It is also worth mentioning that NSE data has been successfully interpreted using a logarithmic form of the normalised intermediate scattering function,

$$\frac{I(Q,t)}{I(Q,t=0)} = A - B\ln\left(\frac{t}{\tau}\right) \qquad (8.16)$$

As an example, and as reported by Arbe et al.,[23] eqn (8.16) has been used to model the temporal evolution of correlations between the structural units within nano-domains in poly(n-alkyl methacrylates) i.e. side group structural relaxation.

8.4   A Word About Water

No matter the research theme, an experimentally determined translational diffusion coefficient, D, from an aqueous system will most likely be compared to that of bulk water at room temperature ($D_{H_2O,QENS,293K}$ ~ 2.02×10^{-9} m^2 s^{-1}).[26] As a result, data collection from bulk H_2O, and subsequent modelling, accompanies many a QENS investigation as a reference measurement. Two analysis hypotheses developed to model QENS spectra from bulk water are presented below.

The *standard model* used to describe bulk water diffusion is that of Teixeira et al.,[7] which built on the work of Chen et al.[27] The standard model assumes that bulk water is a system in which proton dynamics are assigned to either continuous localised diffusion or jump-like translational states; the two states being assumed independent. It was proposed that these states could be decoupled by adjusting temperature, with measurements extending into a super-cooled water regime being attributed to, and modelled in terms of, molecular rotations alone. The Teixeira *et al.* approach has been widely adopted and has underpinned many aqueous solution studies.[28,29]

In terms of the self-intermediate scattering function, $I_s(\boldsymbol{Q},t)$, the Teixeira model is written,[7]

$$I_s(\boldsymbol{Q},t) = \exp\left(-\frac{Q^2 \langle u^2(T)\rangle}{3}\right) R(\boldsymbol{Q},t) T(\boldsymbol{Q},t) \tag{8.17}$$

$R(\boldsymbol{Q},t)$ represents the low-frequency rotations of the molecule. $T(\boldsymbol{Q}, t)$ represents translational motion. Here,

$$R(Q,t) = j_0^2(Qa) + 3j_1^2(Qa)\exp\left(-\frac{t}{3\tau_1}\right) + 5j_2^2(Qa)\exp\left(-\frac{t}{\tau_1}\right) \tag{8.18}$$

and,

$$T(Q, t) = \exp(-\Gamma(Q)t) \tag{8.19}$$

a is the radius of rotation, which for the water molecule is the O–H distance. τ_1 is the relaxation time associated with the localised diffusion. j_n are the spherical Bessel functions associated with the Sears[30] expression for the classical localised diffusion of a molecule. $\Gamma(Q)$ is given by model No. 5 in Table 8.1.

However, as Teixeira *et al.* readily acknowledged, the assumptions used to develop the model needed validation. Not only were there discrepancies between rotational correlation times derived using the standard model and those reported from other more direct methods,[31] but the pronounced deviation from the continuous diffusion limit, DQ^2, at low temperatures and high Q observed in the experimental data was not as distinct in simulation. Other attempts to interpret the localised contribution to the quasi-elastic neutron scattering spectra from super-cooled water included the relaxed cage model (RCM);[32] an approach in which the simple exponential translational and localised relaxation functions used in the standard model were replaced by stretched exponentials. While providing a good fit to the data, the RCM was difficult to test and, for some applications, required rotation–translation decoupling and the elastic incoherent structure factor approximations to be retained.

In 2011, Qvist *et al.*[26] proposed a new interpretation for the bulk water problem based on the dynamical clustering observed from simulations. Using a model free approach, and by benchmarking the experimental results against detailed molecular dynamics simulation, Qvist *et al.* described the QENS data in terms of two motional components that aligned with two distinct types of structural dynamics: picosecond local structural fluctuations within dynamical basins and slower inter-basin jumps. The fit model used was,

$$I(Q, \omega; \lambda_0) = C(Q; \lambda_0)\{A(Q)v_1(Q, \omega; \lambda_0) + [1 - A(Q)] \times v_2(Q, \omega; \lambda_0)\} + B(Q; \lambda_0) \quad (8.20)$$

where $A(Q) = \cos^2\theta(Q)$. The Voigt function, $v_n(Q, \omega; \lambda_0)$ is the convolution of the neutron instrument's resolution function and a normalised Lorentzian function (HWHM, Γ_n),

$$L_n(Q,\omega) = \frac{1}{\pi} \frac{\Gamma_n(Q)}{[\Gamma_n(Q)]^2 + \omega^2} \quad (8.21)$$

with the result of this convolution being expressed as,

$$v_n(Q, \omega; \lambda_0) \equiv L_n(Q, \omega) \otimes R(Q, \omega; \lambda_0)$$

$$= \frac{1}{(\sqrt{2\pi})\sigma(Q;\lambda_0)} \Re\left[\exp(-Z_n^2)\mathrm{erfc}(-iZ_n)\right] \quad (8.22)$$

where $Z_n \equiv (\omega + i\Gamma_n)/((\sqrt{2})\sigma)$.

While the *standard model* is used routinely, and coded into many analysis packages, the Qvist's approach better associates with other experimental methods; the macroscopic diffusion coefficient deduced from the QENS data being found to agree quantitatively with NMR $(D_{\mathrm{QENS, 293\ K}} = 2.017(7) \times 10^{-9}\ \mathrm{m^2\ s^{-1}}, D_{\mathrm{NMR, 293\ K}} = 2.04 \times 10^{-9}\ \mathrm{m^2\ s^{-1}})$. Nonetheless, as the evolution of the bulk water diffusion problem highlights, the need for continuous development of neutron instrumentation, coupled with advances in computational methods, is key if scientific understanding is to evolve.

8.5 Further Reading

By presenting a general overview of the measurement and analysis of quasi-elastic neutron scattering data, the research examples given, and models presented, in this book are clearly expansive. It is the hope that the references contained within might focus more specifically on your particular research interest. However, it is worth concluding by guiding you toward alternative, yet complementary, resources.

The works listed below fall into two main categories: (i) edited subject overviews, the content of which is science rather than method focused and (ii) authored technique focused tomes. For those interested in the former, (i), you are directed towards:

- *Dynamics of Soft Matter: Neutron Applications*, ed. Victoria García Sakai, Christiane Alba-Simionesco and Sow Hsin Chen, 2012.[33]
- *Neutron Scattering in Biology*, ed. J. Fitter, T. Gutberlet and J. Katsaras, 2006.[34]
- *Quasi-elastic Neutron Scattering for the Diffusive Motions in Solids and Liquids*, T. Springer, 1972.[35]
- *Dynamics of Biological Macromolecules by Neutron Scattering*, ed. S. Magazù and F. Migliardo, 2011.[36]
- *Quasi-elastic Neutron Scattering and Solid State Diffusion*, R. Hempelmann, 2000.[37]
- *Polymers and Neutron Scattering*, Julia S. Higgins and Henri C. Benoît, 1996.[38]
- *Applications of Neutron Scattering to Soft Condensed Matter*, ed. B. J. Gabrys.[39]
- *Neutron Spin Echo in Polymer Systems*, D. Richter, M. Monkenbusch, A. Arbe and J. Colmenero, 2005.[40]

For the latter, (ii),

- *Quasi-elastic Neutron Scattering Principles and Application in Solid State Chemistry, Biology and Materials Science*, M. Bée, 1988.[41]
- *Neutron Spin Echo Spectroscopy: basics, trends, and applications*, F. Mezei, C. Pappas and T. Gutberlet, 2003.[42]
- *Neutron Scattering at a Pulsed Source*, ed. R. J. Newport, B.D. Rainford and R. Cywinski, 1988.[43]
- *Pulsed Neutron Scattering*, C. G. Windsor, 1981.[44]
- *Introduction to the Theory of Thermal Neutron Scattering*, G. L. Squires, 1996.[45]

References

1. D. S. Sivia and C. J. Carlile, *J. Chem. Phys.*, 1992, **96**, 170–178.
2. D. S. Sivia, C. J. Carlile, W. S. Howells and S. Konig, *Physica B*, 1992, **182**, 341–348.
3. A. M. Stadler, J. P. Embs, I. Digel and G. M. Artmccai, *J. Am. Chem. Soc.*, 2008, **130**, 16852–16853.
4. K. S. Singwi and A. Sjölander, *Phys. Rev.*, 1960, **120**, 1093–1102.
5. M. Bée, *Quasi-elastic Neutron Scattering: Principles and Applications in Soild State Chemistry, Biology and Materials Science*, Adam Hilger, Bristol, England, 1988.
6. A. Fick, *J. Membr. Sci.*, 1995, **100**, 33–38.

7. J. Teixeira, M. C. Bellissent-Funel, S. H. Chen and A. J. Dianoux, *Phys. Rev. A*, 1985, **31**, 1913–1917.
8. G. Paradossi, M. T. Di Bari, M. T. F. Telling, A. Turtu and F. Cavalieri, *Physica B*, 2001, **301**, 150–156.
9. C. T. Chudley and R. J. Elliott, *Proc. Phys. Soc.*, 1961, **77**, 353–361.
10. P. L. Hall and D. K. Ross, *Mol. Phys.*, 1981, **42**, 673–682.
11. A. J. O'Malley, M. Sarwar, J. Armstrong, C. R. A. Catlow, I. P. Silverwood, A. P. E. York and I. Hitchcock, *Phys. Chem. Chem. Phys.*, 2018, **20**, 11976–11986.
12. J. H. Roh, J. E. Curtis, S. Azzam, V. N. Novikov, I. Peral, Z. Chowdhuri, R. B. Gregory and A. P. Sokolov, *Biophys. J.*, 2006, **91**, 2573–2588.
13. S. Khodadadi, J. H. Roh, A. Kisliuk, E. Mamontov, M. Tyagi, S. A. Woodson, R. M. Briber and A. P. Sokolov, *Biophys. J.*, 2010, **98**, 1321–1326.
14. L. Hong, N. Smolin, B. Lindner, A. P. Sokolov and J. C. Smith, *Phys. Rev. Lett.*, 2011, **107**, 148102.
15. A. Arbe and J. Colmenero, *Phys. Rev. E*, 2009, **80**, 041805.
16. J. Tanchawanich, V. Arrighi, M. C. Sacchi, M. T. F. Telling and A. Triolo, *Macromolecules*, 2008, **41**, 1560–1564.
17. W. S. Howells, *Phys. B*, 1996, **226**, 78–81.
18. B. J. Gold, W. Pyckhout-Hintzen, A. Wischnewski, A. Radulescu, M. Monkenbusch, J. Allgaier, I. Hoffmann, D. Parisi, D. Vlassopoulos and D. Richter, *Phys. Rev. Lett.*, 2019, **122**, 088001.
19. D. Richter, M. Monkenbusch, A. Arbe and J. Colmenero, in *Neutron Spin Echo in Polymer Systems*, ed. D. Richter, M. Monkenbusch, A. Arbe and J. Colmenero, Springer Berlin Heidelberg, Berlin, Heidelberg, 2005, pp. 1–221.
20. E. G. Kelley, P. D. Butler and M. Nagao, *Soft Matter*, 2019, **15**, 2762–2767.
21. K. Weron, *J. Phys.: Condens. Matter*, 1991, **3**, 9151–9162.
22. A. G. Zilman and R. Granek, *Phys. Rev. Lett.*, 1996, **77**, 4788–4791.
23. A. Arbe, A. C. Genix, S. Arrese-Igor, J. Colmenero and D. Richter, *Macromolecules*, 2010, **43**, 3107–3119.
24. K. Wood, C. Caronna, P. Fouquet, W. Haussler, F. Natali, J. Ollivier, A. Orecchini, M. Plazanet and G. Zaccai, *Chem. Phys.*, 2008, **345**, 305–314.
25. R. M. Pickup, R. Cywinski, C. Pappas, B. Farago and P. Fouquet, *Phys. Rev. Lett.*, 2009, **102**, 097202.
26. J. Qvist, H. Schober and B. Halle, *J. Chem. Phys.*, 2011, **134**, 144508.
27. S. H. Chen, J. Teixeira and R. Nicklow, *Phys. Rev. A*, 1982, **26**, 3477–3482.
28. F. Cavatorta, A. Deriu, D. Dicola and H. D. Middendorf, *J. Phys.: Condens. Matter*, 1994, **6**, A113–A117.
29. M. R. Harpham, N. E. Levinger and B. M. Ladanyi, *J. Phys. Chem. B*, 2008, **112**, 283–293.
30. V. F. Sears, *Can. J. Phys.*, 1966, **44**, 1279–1297.
31. J. Teixeira, M. C. Bellissentfunel and S. H. Chen, *J. Mol. Liq.*, 1991, **48**, 123–128.
32. L. Liu, A. Faraone and S.-H. Chen, *Phys. Rev. E*, 2002, **65**, 041506.
33. V. Garcia Sakai, C. Alba-Simionesco and S. H. Chen, *Dynamics of Soft Matter: Neutron Applications*, Springer US, 2011.
34. J. Fitter, T. Gutberlet and J. Katsaras, *Neutron Scattering in Biology: Techniques and Applications*, Springer, Berlin; London, 2011.
35. T. Springer, *Quasielastic Neutron Scattering for the Investigation of Diffusive Motions in Solids and Liquids*, Springer, 1972.
36. F. Migliardo and S. Magazu, *Dynamics of Biological Macromolecules by Neutron Scattering*, Bentham eBooks, 2011.
37. R. Hempelmann, *Quasielastic Neutron Scattering and Solid State Diffusion*, Oxford University Clarendon Press, 2000.
38. J. S. Higgins, H. C. Benoît and H. Benoît, *Polymers and Neutron Scattering*, Clarendon Press, 1996.

39. B. J. Gabrys, *Applications of Neutron Scattering to Soft Condensed Matter*, Gordon and Breach Science Publishers, 2014.
40. D. Richter, *Neutron Spin Echo in Polymer Systems*, Springer, Berlin, 2005.
41. M. Bée, *Quasielastic Neutron Scattering*, Adam Hilger, 1988.
42. F. Mezei, C. Pappas and T. Gutberlet, *Neutron Spin Echo Spectroscopy: Basics, Trends and Applications*, Springer, Berlin; New York, 2003.
43. R. J. Newport, B. D. Rainford and R. Cywinski, *Neutron Scattering at a Pulsed Source*, A. Hilger, Bristol, England; Philadelphia, PA, USA, 1988.
44. C. G. Windsor, *Pulsed Neutron Scattering*, Taylor and Francis, London, 1981.
45. G. L. Squires, *Introduction to the Theory of Thermal Neutron Scattering*, Dover Publications, 1996.

9   And Finally

At its most fundamental, neutron scattering allows the research community to ask: *'where atoms are'* and *'what atoms do'*. Answers are possible because of the unique properties of this highly penetrating, weakly perturbing, yet non-destructive, probe. Not only do neutrons possess wavelengths that span inter-atomic distances and have energies comparable to vibrational frequencies and diffusive motions, but they offer the possibility to label with deuterium multi component materials and thus unravel the characteristics of highly complex, hydrogenous systems.

This specialised 'primer style' training resource has hopefully convinced you that the technique of quasi-elastic neutron scattering in particular is a powerful research tool for extracting temporal and spatial information at the nanoscale from both soft and hard condensed matter. It is also a hope that the content will enhance research success in your field of study, or perhaps illuminate previously unforeseen possibility.

Before we conclude, however, a comment should be made concerning the interplay of QENS and molecular dynamics (MD) simulations. While a topic beyond the scope of this book, it is clear from the case studies presented that the use of MD to support neutron experiments is prevalent, with the temporal and spatial ranges accessible using, and parameters extracted from, MD aligning directly with those probed experimentally. The QENS method is thus an excellent means with which to validate or advance MD models. Conversely, MD simulations allow for the planning and optimisation of QENS experiments either *via* simulation of the scattering function alone or in conjunction

A Practical Guide to Quasi-elastic Neutron Scattering
By Mark T. F. Telling
© Mark T. F. Telling 2020
Published by the Royal Society of Chemistry, www.rsc.org

with virtual neutron instrument scattering investigations.[1] It is worth highlighting therefore that concerted effort is focused not only on the expansion of neutron facility instrumentation, but also the development, improvement and support of software tools and protocols that make the extraction of neutron specific parameters from MD simulation trajectory files routine, especially for the non-specialist lay-user.[2,3]

For now, however, let the real challenge begin; interpreting the parameters that the quasi-elastic neutron scattering method has helped you extract!

The very best of luck.

References

1. E. Farhi, V. Hugouvieux, M. R. Johnson and W. Kob, *J. Comput. Phys.*, 2009, **228**, 5251–5261.
2. G. Goret, B. Aoun and E. Pellegrini, *J. Chem. Inf. Model.*, 2017, **57**, 1–5.
3. K. Hinsen, E. Pellegrini, S. Stachura and G. R. Kneller, *J. Comput. Chem.*, 2012, **33**, 2043–2048.

Appendix 1
Conversion Factors

Table A1.1 Key conversion factors and equations to help you align the units reported in this book with those extracted using other experimental methods. Table to be read left to right; for example, 1 Kelvin (K) equals 1.99×10^{-3} kcal mol^{-1}. Adapted from ref. 1 with permission from Taylor and Francis, Copyright 1988.[a]

	kJ mol^{-1}	kcal mol^{-1}	meV	cm^{-1}	rad s^{-1}	Hz	K	Å	m s^{-1}
kJ mol^{-1}	1	0.239	10.36	83.58	1.57×10^{13}	2.51×10^{12}	120.3	2.81	1.41×10^{3}
kcal mol^{-1}	4.19	1	43.39	345	6.59×10^{13}	1.05×10^{13}	503.3	1.373	2.88×10^{3}
meV	0.097	0.023	1	8.01	1.52×10^{12}	2.42×10^{11}	11.61	9.05	437.4
cm^{-1}	1.19×10^{-2}	2.86×10^{-3}	0.124	1	1.88×10^{11}	2.99×10^{10}	1.439	25.68	154.05
rad s^{-1}	6.35×10^{-14}	1.52×10^{-14}	6.58×10^{-13}	5.31×10^{-12}	1	0.16	7.64×10^{-12}	1.12×10^{7}	3.55×10^{-4}
Hz	3.99×10^{-13}	9.54×10^{-14}	4.14×10^{-14}	3.34×10^{-11}	6.238	1	4.8×10^{-11}	4.45×10^{6}	8.89×10^{-4}
K	8.31×10^{-3}	1.99×10^{-3}	8.62×10^{-2}	0.695	1.31×10^{11}	2.08×10^{10}	1	30.81	128.4
Å	7.894	1.89	81.807	659.8	1.24×10^{14}	1.98×10^{13}	949.4	1	3.96×10^{3}
m s^{-1}	5.04×10^{-7}	1.2×10^{-7}	5.23×10^{-6}	4.22×10^{-5}	7.95×10^{6}	1.27×10^{6}	6.07×10^{-5}	3.96×10^{3}	1

[a]kcal, kJ, meV - units associated with energy, E; cm^{-1} - unit associated with the optical wave vector; rad s^{-1} - unit associated with cycles, ω, where $E = \hbar\omega$; Hz - unit associated with frequency, $\nu = \omega/2\pi$; K - unit associated with temperature, T, where $E = k_B T$; Å - unit associated with wavelength, λ, where $E = \hbar^2/2m_n\lambda^2$ and $\lambda \sim 1/\sqrt{E}$; m s^{-1} - unit associated with velocity, v, where $E = m_n v^2/2$ and $v \sim \sqrt{E}$; $\hbar$ - reduced Planck constant ($\hbar = h/2\pi$) 1.054×10^{-34} Js; m_n - mass of the neutron, 1.6749×10^{-27} kg.

A Practical Guide to Quasi-elastic Neutron Scattering
By Mark T. F. Telling
© Mark T. F. Telling 2020
Published by the Royal Society of Chemistry, www.rsc.org

Reference

1. M. Bée, *Quasi-elastic Neutron Scattering: Principles and Applications in Soild State Chemistry, Biology and Materials Science*, Adam Hilger, Bristol, England, 1988.

Appendix 2
Supporting Measurements

Table A2.1 The pre- and post-experimental methods referred to in the Chapter 1 case studies with a short description of the supporting information each technique affords.

Technique	Use
Nuclear magnetic resonance (NMR)	Determine activated 4-cyanopentanoic acid dithiobenzoate (CPDB) *via* ^{1}H-NMR[1]
Differential scanning calorimetry (DSC)	Identify glass (T_g) transition temperatures[2] Identify endothermic and exothermic transitions[3]
Thermo-gravimetric analysis (TGA)	Characterise the percentage of structural (bound) hydrogen[4]
Fourier-transform infrared spectroscopy (FTIR)	Investigate structural changes in the Amyloid β (Aβ) peptide induced by the DMPG lipid bilayers[5]
X-ray powder diffraction (XRD)	Locate angular positions of structural Bragg lines (possible QENS signal contaminants) from materials exhibiting long range crystallographic order. Check sample phase purity[6]
Sodium dodecyl sulfate polyacrylamide gel electrophoresis (SDS-PAGE)	Ensure biological sample purity *via* size separation analysis[7]

A Practical Guide to Quasi-elastic Neutron Scattering
By Mark T. F. Telling
© Mark T. F. Telling 2020
Published by the Royal Society of Chemistry, www.rsc.org

Table A2.1 (continued)

Technique	Use
Small angle neutron/X-ray scattering (SANS/SAXS)	Measure effect of high hydrostatic pressure on the structural features of LDL particles[7]
	Structural investigation to explore coexistence of polybutadiene rubber regions with low and high crosslinking densities[2]
	Directly probe the average conformation of the grafted PMA chains in the CPB and SDPB regions[1]
	Expose structures at different locations and over different length scales *via* size and scattering length density changes[5]
Dynamic viscoelastic measurement	Extract a relationship between the elastic modulus and the volume fraction of zinc diacrylate[2]
Brunauer–Emmett–Teller nitrogen adsorption isotherms (BET method)	Identify the specific surface area of the QENS samples[8]
Molecular dynamics (MD) simulation	Residue resolved interpretation of the experimental neutron scattering data[9]
	Simulation of the self-diffusion coefficient of ammonia[10]
	Allow comparison of theoretical and experimental scattering functions for a single benzene molecule on a graphite surface[8]
Size exclusion chromatography with multi-angle light scattering (SEC-MALS)	Characterise the molecular weight, and dispersity, of synthesised polymers[1]
Dynamic light scattering (DLS)	Extract bulk diffusion coefficients[9]
	A combination of dynamic light scattering (DLS) and depolarised DLS to allow access to translational and rotational diffusion coefficients of bulk BPM, salol molecules and those in a dissolved state[11]
	Get initial structural parameters, such as polydispersity, for SANS data fitting[5]
Ultraviolet–visible spectroscopy (UV-Vis)	Determine percentage of grafted 4-cyanopentanoic acid dithiobenzoate (CPDB)[1]
	Investigate light-dependent absorption changes in the visible region[9]
Dielectric spectroscopy (DS)	Access to the dielectric properties of the ionic liquids, *i.e.* the frequency dependence of the dielectric constant, dielectric loss and the real part of the conductivity[3]

(*continued*)

Table A2.1   (continued)

Technique	Use
AC susceptibility (χ_{ac})	Characterise magnetic ground states and identify spin glass transition temperatures and frequency response[6]
Inelastic neutron spectroscopy (INS)	Probe changes in the vibrational response as a function of cement age thus gain insight into the molecular properties, and development, of the hydrogen-bond network[4]
Circular dichroism (CD)	Determine the secondary structure of the Amyloid β (Aβ) peptide and ascertain the optimum concentration at which a significant portion of Aβ binds with the DMPG lipid bilayer[5]
Neutron membrane diffraction (NMD)	Resolve details of the Amyloid β (Aβ) peptide in the DMPG lipid bilayers[5]

References

1. Y. Wei, Y. F. Xu, A. Faraone and M. J. A. Hore, *ACS Macro Lett.*, 2018, **7**, 699–704.
2. R. Mashita, R. Inoue, T. Tominaga, K. Shibata, H. Kishimoto and T. Kanaya, *Soft Matter*, 2017, **13**, 7862–7869.
3. C. J. Jafta, C. Bridges, L. Haupt, C. Do, P. Sippel, M. J. Cochran, S. Krohns, M. Ohl, A. Loidl, E. Mamontov, P. Lunkenheimer, S. Dai and X. G. Sun, *Chemsuschem*, 2018, **11**, 3512–3523.
4. M. C. Berg, A. R. Benetti, M. T. F. Telling, T. Seydel, D. H. Yu, L. L. Daemen and H. N. Bordallo, *ACS Appl. Mater. Interfaces*, 2018, **10**, 9904–9915.
5. D. K. Rai, V. K. Sharma, D. Anunciado, H. O'Neill, E. Mamontov, V. Urban, W. T. Heller and S. Qian, *Sci. Rep.*, 2016, **6**, 30983.
6. R. M. Pickup, R. Cywinski, C. Pappas, B. Farago and P. Fouquet, *Phys. Rev. Lett.*, 2009, **102**, 097202.
7. M. Golub, B. Lehofer, N. Martinez, J. Ollivier, J. Kohlbrecher, R. Prassl and J. Peters, *Sci. Rep.*, 2017, 7, 46034.
8. I. Calvo-Almazan, E. Bahn, M. M. Koza, M. Zbiri, M. Maccarini, M. T. F. Telling, S. Miret-Artes and P. Fouquet, *Carbon*, 2014, **79**, 183–191.
9. A. M. Stadler, E. Knieps-Grunhagen, M. Bocola, W. Lohstroh, M. Zamponi and U. Krauss, *Biophys. J.*, 2016, **110**, 1064–1074.
10. A. J. O'Malley, I. Hitchcock, M. Sarwar, I. P. Silverwood, S. Hindocha, C. R. A. Catlow, A. P. E. York and P. J. Collier, *Phys. Chem. Chem. Phys.*, 2016, **18**, 17159–17168.
11. C. J. Chen, R. P. Krishnan, K. K. Wong, D. H. Yu, F. Juranyi and S. M. Chathoth, *Phys. Rev. B*, 2018, **98**, 094203.

Appendix 3
Sample Cell 'Background' Features

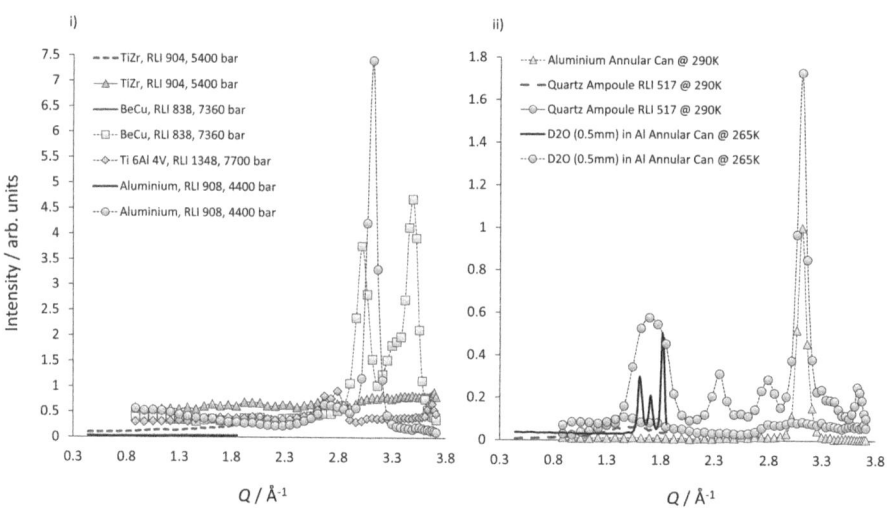

Figure A3.1 Elastic channel intensity *vs.* Q from (i) empty cylindrical pressure cells and (ii) standard aluminum (annular) and quartz (ampoule) sample cells, as well as ice peak positions expected from D_2O cooled to 265 K in said annular container. Data collected on the indirect geometry spectrometer, IRIS, using two instrument configurations that probe momentum transfer ranges (a) 0.2 < Q < 1.8 Å^{-1} (solid and dashed lines) and (b) 0.8 < Q < 3.7 Å^{-1} (filled symbols). Each data set has been normalised to the incident beam monitor and corrected for detector efficiency to allow relative scattering intensities between the different cells to be compared. For reference, the ISIS facility pressure cell codes are given, *i.e.* the BeCu cell measured was cell RLI 838. Upper working pressures at 20 °C are quoted. Note that instrument configuration, (a), requires a cooled beryllium filter to be placed between the sample position and final energy analyser which, as seen, greatly reduces background scatter from other sources.

A Practical Guide to Quasi-elastic Neutron Scattering
By Mark T. F. Telling
© Mark T. F. Telling 2020
Published by the Royal Society of Chemistry, www.rsc.org

Subject Index

Page numbers in *italics* refer to a figure. Page numbers with a suffix T indicate a table.